A PLUME BOOK

EARTHTALK

The E – THE ENVIRONMENTAL MAGAZINE editorial team includes: executive editor and publisher Doug Moss, who founded *E* and created *EarthTalk*; editor Brita Belli, formerly with Connecticut's *Fairfield County Weekly* newspaper; Roddy Scheer, editor at large and *EarthTalk*'s primary writer; and Jim Motavalli, former *E* editor and now senior writer with the magazine.

Also from *E – The Environmental Magazine*

Green Living

EARTH TALK

Expert Answers to Everyday Questions About the Environment

Selections from

E – The Environmental Magazine's

nationally syndicated column

A PLUME BOOK

PLUME
Published by the Penguin Group
Penguin Group (USA) Inc., 375 Hudson Street, New York, New York 10014, U.S.A.
Penguin Group (Canada), 90 Eglinton Avenue East, Suite 700, Toronto,
Ontario, Canada M4P 2Y3 (a division of Pearson Penguin Canada Inc.)
Penguin Books Ltd., 80 Strand, London WC2R 0RL, England
Penguin Ireland, 25 St. Stephen's Green, Dublin 2, Ireland (a division of Penguin Books Ltd.)
Penguin Group (Australia), 250 Camberwell Road, Camberwell, Victoria 3124,
Australia (a division of Pearson Australia Group Pty. Ltd.)
Penguin Books India Pvt. Ltd., 11 Community Centre,
Panchsheel Park, New Delhi - 110 017, India
Penguin Group (NZ), 67 Apollo Drive, Rosedale, North Shore 0632,
New Zealand (a division of Pearson New Zealand Ltd.)
Penguin Books (South Africa) (Pty.) Ltd., 24 Sturdee Avenue,
Rosebank, Johannesburg 2196, South Africa

Penguin Books Ltd., Registered Offices: 80 Strand, London WC2R 0RL, England

First published by Plume, a member of Penguin Group (USA) Inc.

First printing, March 2009
1 3 5 7 9 10 8 6 4 2

LIBRARY OF CONGRESS CATALOGING-IN-PUBLICATION DATA
Earthtalk : expert answers to everyday questions about the environment :
selections from E/the environmental magazine's nationally syndicated column.
p. cm.
Includes index.
ISBN 978-0-452-29012-9
1. Environmental sciences—Miscellanea. 2. Environmental sciences—Popular works.
3. Environmental protection—Citizen participation—Miscellanea. 4. Environmental protection—Citizen participation—Popular works. 5. Green living—Miscellanea. 6. Green living—Popular works. I. E (Norwalk, Conn.)
GE110.E37 2009
333.72—dc22 2008030320

Printed in the United States of America
Set in Legacy Serif Book
Designed by Level C

For the loyal readers of *E – The Environmental Magazine* and "EarthTalk" who have contributed such engaging and informed questions over the years.

CONTENTS

PREFACE

If there was any doubt that Americans care about our beleaguered Earth, it was erased by the results of a 2005 Harris Poll, which found that three-quarters of all Americans agree that the environment should be protected "regardless of cost." A Duke University poll that same year found a whopping 79 percent favoring "stronger national standards to protect our land, air, and water."

We agree that the planet needs our help, but we disagree on what to do. What's more important, organic or local food (or both)? Replacing my energy-hogging refrigerator or my gas-guzzling car? Bagging my groceries in paper or plastic (or neither and bringing my own cloth bag)? Wrapping my baby in cloth diapers or disposable? Washing my dishes by hand or in a dishwasher?

Since our column "EarthTalk" was launched in 1996, we here at *E – The Environmental Magazine* have been fielding a mailbag full of readers' lively questions. At first, we ran the column just in *E* and on our website at www.emagazine.com, answering three questions per bimonthly issue. But we were receiving at least that many burning questions every week and soon realized we had the makings of a column with mainstream appeal outside *E*'s own readership.

In 2000, we signed with a major news syndicate, and the column began running weekly in twenty-one daily newspapers, including the *Houston Chronicle, Minneapolis Star Tribune, Seattle Post-Intelligencer, Sacramento Bee, New Jersey Star-Ledger,* and *Detroit News.*

Twenty-one newspapers was nice coverage, but we wanted more, so in 2003 we took a radical step and decided to syndicate the column ourselves. We also knew it would be an easier sell if it didn't cost anything, so we offered it *without charge* to newspapers, magazines, and websites around the United States and around the world.

Well, the idea worked spectacularly well. Signing up for "EarthTalk" remains free, and some sixteen hundred media outlets, large and small, are now getting the column every week, reaching as many as sixty million readers.

We're very proud that "EarthTalk" is reaching a wide, mainstream audience. And now it's a book! We hope you'll use this collection as a handy reference guide to the green life in an increasingly complex world.

We firmly believe that climate change, peak oil, and water shortages will soon make it plain that we're living in the "environmental century," and objective, scientifically based information arms us all for the challenges ahead.

—The Editors

ACKNOWLEDGMENTS

The editors and writers who contributed to this book are grateful for the support of their families, friends, and colleagues.

The book's primary author, Roddy Scheer, thanks his family for putting up with him working on "EarthTalk" answers day and night.

Doug Moss, the founder of "EarthTalk," thanks his wife, Deborah, and sons, Tim and Jeff, for tolerating his "workaholic" tendencies in the service of *E Magazine* and this column that grew from it. He also wishes to acknowledge the support and trust of the more than seventeen hundred newspaper, magazine, and website editors who receive the column's weekly installments and manage to find space for it in their publications.

Jim Motavalli thanks his nature girls, Mary Ann, Delia, and Maya. Brita Belli thanks the staff at *E* for showing her the green light, and Jerry and Elli for finding joy in hidden places.

Last (but not least), we wish to thank the private foundations that have contributed to this nonprofit effort since its launch in 2003: the Changing Horizons Fund, Colcom Foundation, Geraldine R. Dodge Foundation, Educational Foundation of America, Marisla Foundation, Curtis and Edith Munson Foundation, Park Foundation, V. Kann Rasmussen Foundation, and Sherman Foundation.

Got an environmental question that you don't see answered in this book? Send it to: "EarthTalk," c/o *E – The Environmental Magazine*, P.O. Box 5098, Westport, CT 06881. You can also submit your question at www.emagazine.com or e-mail us at earthtalk@emagazine.com.

EarthTalk

Expert Answers to

Everyday Questions

About the Environment

1

EAT, DRINK, AND BE WARY

From Local Food to Lunchables

Nothing's more personal than what we put in our mouths. We're talking about the food and drink that we hold dear: from that morning pot of coffee and the after-work beer to the stadium chili dog and the high-fat-and-sodium Lunchables that our kids just have to, *have to* (insert high-pitched whine here) have. The American dietary landscape is tough to navigate, full of pitfalls like aspartame-loaded soda and preservative-laden snacks. Sure, we're making progress, but it's a jungle out there. As if the grocery store aisles aren't confusing enough with their coded food package messages reading "all natural" or "cholesterol free," we're now being advised not just to eat our vegetables but to make sure they're locally grown and organic (and preferably raw). In fact, we should just go plant a garden, hoe it, weed it, wait a few months for the vegetables to grow, eat as many as possible, and can the rest for maximum nutritional benefit. What's so hard about that? But even without that kind of commitment, there's quite a bit you can do to eat better and feel better about what you're eating. Seriously, we're here to help you enjoy what's on your plate.

I've heard some environmental advocates claim that organic farming could produce enough food to feed the whole world. Is this true?

Gabe Morello, Lynnwood, WA

There's no doubt that heavy pesticides and single-crop plantings (known as "monoculture") produce a lot of crops while exacting an ugly environmental toll. Evidence from the United States, Britain, and

Australia shows that such high-tech fields produce as much as 40 percent more crops than the old-fashioned manure methods.

But organic farming can top those numbers. According to articles in the journal *Nature*, organic methods for growing rice, corn, and wheat all produced significantly higher yields—at less cost—than monoculture farms. Researchers at the University of Michigan (UM) found that organic farming methods could produce equal yields at home and double or even triple yields in developing countries. "My hope is that we can finally put a nail in the coffin of the idea that you can't produce enough food through organic agriculture," said Ivette Perfecto, a professor at UM's School of Natural Resources and Environment.

Farmers in India, Kenya, Brazil, Guatemala, and Honduras have doubled or tripled their yields by switching to organic agriculture. Cuban farmers, who cannot access fertilizers and pesticides due to the U.S. embargo, have also seen higher yields by going organic.

And organic farming doesn't punish the land. "Organic farming methods continually increase soil fertility and prevent loss of topsoil to erosion," says Dr. Christos Vasilikiotis of the University of California, Berkeley. "Conventional methods have the opposite effect."

Even when they don't get a bumper crop, organic farmers make up the financial difference by not having to buy costly pesticides and fertilizers. And scientists say that improved growing techniques and new natural pest controls could eventually level the playing field, giving organic farmers the economic advantage.

Only slightly more than 2 percent of all farms in the United States are currently organic. But with sales of domestic organic food growing about 20 percent annually, that figure is expected to rise exponentially in years to come.

Still, feeding the world is a tall order, and everyone from organic farmers to environmental leaders to human rights workers agrees that ending hunger is dependent more upon political will than agricultural prowess. "Until governments tackle the social and political factors involved in poverty and food distribution, millions of people will continue to go hungry," concludes Dr. Liz Stockdale of Britain's Institute of Arable Crops Research.

CONTACTS: Department of Agriculture, National Organic Program, www.ams.usda.gov/nop; Organic Consumers Association, www.organicconsumers.org.

Why do environmentalists advocate that people "eat locally"? I don't understand the connection between patronizing local food producers and environmental quality.

Timothy Douglas, Burlington, VT

In this age of food preservatives and additives, genetically altered crops, and *E. coli* outbreaks in spinach, people are increasingly concerned about the quality and cleanliness of the foods they eat. Given the impossibility of identifying the pesticides used to grow and the route taken to transport, say, a banana from Central America to our local supermarket, foods grown locally make a lot of sense for those who want more control over what they put into their bodies.

John Ikerd, a retired agricultural economics professor who writes about the "eat local" movement, says that farmers who sell direct to consumers need not give priority to packing, shipping, and shelf-life issues and can instead "select, grow, and harvest crops to ensure peak qualities of freshness, nutrition, and taste." Eating local also means eating seasonally, he adds, a practice much in tune with Mother Nature.

"Local food is often safer, too," says the Center for a New American Dream (CNAD). "Even when they are not organic, small farms tend to be less aggressive than large factory farms about dousing their wares with chemicals." Small farms are also more likely to grow a variety of crops, says CNAD, protecting biodiversity and preserving a wider agricultural gene pool, an important factor in long-term food security.

Eating locally grown food even helps in the fight against global warming. Rich Pirog of the Leopold Center for Sustainable Agriculture reports that the average fresh food item on our dinner table travels fifteen hundred miles to get there. Buying from the neighborhood farm stand eliminates the need for all that fuel-guzzling transportation.

And those fresh tomatoes and peppers support the local economy. Farmers on average receive only twenty cents of each food dollar spent, says Ikerd, the rest going for transportation, processing, packaging, refrigeration, and marketing. Farmers who sell food to local customers "receive the full retail value, a dollar for each food dollar spent," he says. Additionally, eating locally encourages the use of local farmland for farming, thus keeping development in check and preserving open space.

Portland, Oregon's Ecotrust has launched a campaign, the Eat Local Challenge, to encourage people to eat locally for a week so they can see–and taste–the benefits. The organization provides an "Eat Local Scorecard" to those willing to try. Participants must commit to spending 10 percent of their grocery budget on foods grown within a hundred-mile radius of their home. In addition they are asked to try one new fruit or vegetable each day, and to freeze or otherwise preserve some food to enjoy later in the year.

Ecotrust also provides consumers with tips on how to eat locally more often. Shopping regularly at local farmers' markets or farm stands tops the list. Also, locally owned grocery and natural foods stores and co-ops are much more likely than supermarkets to stock local foods. The Local Harvest website provides a comprehensive national directory of farmers' markets, farm stands, and other sources of locally grown food.

CONTACTS: Center for a New American Dream, www.newdream.org/consumer/farmersmarkets.php; Ecotrust Eat Local Challenge, www.eatlocal.net; Local Harvest, www.localharvest.org.

Is it true that some foods we buy contain genetically engineered ingredients that are known to cause health problems?

George Kaye, New York, NY

First made available in the United States during the mid-1990s, genetically modified (GM) foods have snuck into most major American crops while consumers weren't looking. The majority of corn, soy, and cotton grown by American farmers today is from seeds genetically engineered to repel pests without the need for spraying pesticides or herbicides. GM versions of canola, squash, and papaya are also showing up on U.S. shelves.

The potential health effects of GM food products are still unclear. But while conclusive results have been hard to come by, some of the few studies conducted on animals fed diets consisting of GM foods have generated disturbing results.

In one study, potatoes engineered to contain an insect-repelling gene caused intestinal damage in lab mice. While the mice did not die, lesions that formed in their digestive tracts worried researchers enough to recommend more thorough testing of the "transgenic potatoes" before marketing them to humans.

In another study, mice were fed so-called Flavr Savr tomatoes, which were developed in the early 1990s by Calgene and were "optimized for flavor retention." Similar lesions arose in the intestines of the mice, causing reviewers from the U.S. Food and Drug Administration (FDA) to conclude that "the data fell short of 'a demonstration of safety,'" adding that "unresolved questions still remain." Yet later, yielding to the pressure of industry lobbyists, the FDA not only approved the Flavr Savr for mass human consumption but also claimed that all safety issues had been satisfactorily resolved.

When the Flavr Savr hit store shelves, consumers were not particularly impressed with its taste. Farmers also found themselves coping with disease problems and low yields, the very problems the technology sought to address. Eventually, the "Franken tomato," as some cynics dubbed it, was abandoned altogether.

Its legacy lives on, however. Many environmental advocates feel that the Flavr Savr case set the bar particularly low for FDA approval of other GM foods that may or may not cause health problems. Further, it remains to be seen what effects these hybridized species might have on the larger environment, which critics say is reason enough to delay the mass release of GM foods into the market.

Meanwhile, European countries have remained steadfast against allowing GM crops to be grown on their own farms for fear of widespread environmental contamination. And they continue to debate whether or not to allow GM food imports into Europe.

CONTACTS: Belinda Martineau, *First Fruit* (New York: McGraw-Hill, 2001); Pew Initiative on Food and Biotechnology, www.pewagbiotech.org.

What is the environmental impact of sugar, aside from its not-so-healthy aspects? And are there good alternative sweeteners?

Mary Oakes, via e-mail

Our food may be loaded with sugar, from our breakfast cereal to our ketchup, but we rarely think about where the sweet stuff comes from.

According to the World Wildlife Fund (WWF), roughly 145 million tons of sugars are produced in 121 countries each year. And sugar production does indeed take its toll on surrounding soil, water, and air, especially in threatened tropical ecosystems near the equator. Sugar may be responsible for more biodiversity loss than any other crop, the group says, due to the destruction of habitat to make way for plantations, intensive use of water for irrigation, heavy use of agricultural chemicals, and the polluted wastewater that is routinely discharged in the production process.

One extreme example of Big Sugar's environmental damage is the Great Barrier Reef off Australia. Its waters are poisoned by large quantities of effluents, pesticides, and sediment from sugar farms. Further destruction to surrounding wetlands–an integral part of the reef's ecology–also threatens the reef.

In the United States, where sugar production is concentrated in Florida, the health of the Everglades is seriously compromised after decades of sugarcane farming. Tens of thousands of acres have been converted from teeming subtropical forest to lifeless marshland due to excessive fertilizer runoff and drainage for irrigation. A tenuous agreement between environmentalists and sugar producers under the Comprehensive Everglades Restoration Plan has ceded some sugarcane land back to nature and reduced water usage and fertilizer runoff. But it's still uncertain if this and other restoration efforts will bring back Florida's once great "river of grass."

The WWF and other environmental groups are working on public education and legal campaigns to try to reform the international sugar trade. "The world has a growing appetite for sugar," says the WWF's Elizabeth Guttenstein. "Industry, consumers, and policy makers have to work together to make sure that in the future sugar is produced in ways that least harm the environment."

And doctors consider sugar's ubiquity in our society a root cause of numerous health problems, including the rise in obesity and adult-onset diabetes. By far the most commonly used sugar alternative today is aspartame. It's in Diet Coke, Diet Dr Pepper, and other diet sodas; it's in Dentyne and other sugar-free gums; it's the main ingredient in the artificial sweeteners Equal and NutraSweet that Americans stir into their morning coffee. It's hard to think of something so ordinary as "deadly." But aspartame has been linked to a host of health problems, including Parkinson's disease, anxiety attacks, depression, and even brain tumors. A recent federal report listed ninety documented symptoms associated with aspartame exposure. And according to the FDA, aspartame accounts for 75 percent of reported adverse reactions to food additives.

Honey, on the other hand, contains vitamins C, D, E, and B complex, as well as traces of amino acids, enzymes, and minerals. But you're better served by the local variety—up to 50 percent of these nutrients are lost when honey is commercially processed. Also, honey is high in calories and is absorbed by the body in much the same way sugar is, so it's not a good choice if you are diabetic.

Luckily for those with cravings for sweets, several healthy alternatives to sugar do exist and can be found at most natural food markets and increasingly in mainstream supermarkets with natural foods sections. For a honey taste without all the calories, agave nectar—made from the Mexican agave plant—is a good choice because the fruit sugar absorbs more slowly into the bloodstream and is better for diabetics. It has a light, mild flavor with a thinner consistency than honey. One organic brand is Colibree. Another comes from Sweet Cactus Farms.

For baking, date sugar is a good alternative. Made of finely ground dates, it contains all the fruit's nutrients and minerals. Date sugar isn't highly processed, and it can be used cup-for-cup as a replacement for white sugar. Also good for baking is xylitol, which sounds like a chemical but is actually birch sugar. Unlike conventional sugar, xylitol is reported to fight tooth decay, and it has fewer calories. Both date sugar and xylitol are suitable for diabetics and others who are sugar sensitive.

Another sugar alternative—and one whose popularity is rising—is stevia, which comes from the leaf of the stevia plant found in Paraguay. It is about three hundred times as sweet as sugar but has no calories. The FDA considers stevia a dietary supplement, because in its unprocessed form it is very nutritious, containing magnesium, niacin, potassium, and vitamin C. Because stevia is so concentrated, it is best used as an additive to drinks, cereals, or yogurt but not for baking, as it lacks sufficient bulk.

CONTACTS: Colibree, (866) 635-8854 or (720) 272-3969, www.agavenectar.com; Comprehensive Everglades Restoration Plan, www.evergladesplan.org; FDA, (888) INFO-FDA, www.fda.gov; Sweet Cactus Farms, (310) 733-4343, www.sweetcactusfarms.com.

There are many nondairy alternatives to milk and cheese products in my supermarket these days. Are they any good, and what are the health benefits?

Cailin White, San Francisco, CA

Milk does not always do a body good. Dairy products are among the leading causes of food allergies. Many people (a large number of them undiagnosed) are lactose intolerant, and others reject all animal-based foods as part of a strict vegetarian (vegan) diet. Another concern is recombinant bovine growth hormone (rBGH), given to an estimated 30 percent of conventional dairy cows to increase production. Some scientists believe that consuming dairy products from rBGH-treated cows may increase the risk of prostate and premenopausal breast cancer.

Most nondairy offerings come from soy. Soy milk has been around for years, and recently soy milk makers have tweaked tastes and textures to make them more appealing to the American palate. Whether you buy them fresh in the dairy case or in aseptic (paper and foil) packages, you'll notice that taste varies greatly from one brand to the next.

Even nondairy varieties made by the same company can vary, as soy milks come in low-fat, low-carb, vanilla- and chocolate-flavored, unsweetened, and vitamin-fortified versions. Once you find one you like, you can use it cup-for-cup as a milk replacement in most recipes, or just drink it straight. Look for brands labeled "USDA Organic," as those won't contain genetically modified ingredients or residues from pesticides.

Edensoy, one of the original soy milks, is sold in the easily stored aseptic packages, which don't need to be refrigerated until opened. WhiteWave's Silk, sold cold in the dairy sections of many supermarkets, has a smooth taste and even mixes well in coffee. It also comes in vanilla, chocolate, and eggnog flavors. Hain Celestial's Westsoy offers fresh soy shakes and lattes as well as plain milks.

Taste experiences for faux cheese range from delicious to downright inedible. TofuRella cheddar flavor is harsh on the palate, though low in calories and fat compared to its dairy counterpart. Lifetime low-fat Jalapeno Jack, made with organic brown rice, is quite tasty. It is low in fat and calories and tops pizza well (though don't expect it to melt quite as well as conventional cheese).

The very edible and tasty-in-a-sandwich Good Slice Cheddar Style Cheese Alternative from Yves Veggie Cuisine is lower in fat and calories than most others. Galaxy Nutritional Foods offers a particularly wide range of alternative cheeses, from mozzarella to cream cheese to feta crumbles. Dairy-free cheeses are not found as readily as soy milk in mainstream supermarkets but are available in most natural foods markets such as Wild Oats, Mrs. Green's, and others.

CONTACTS: Edensoy, (888) 424-3336, www.edenfoods.com; Galaxy Nutritional Foods, (407) 855-5500, www.galaxyfoods.com; Hain Celestial, (800) 434-4246, www.hain-celestial.com; WhiteWave, (303) 635-4000, www.whitewave.com; Yves Veggie Cuisine, (800) 434-4246, www.yvesveggie.com.

How is the planet affected by our voracious appetite for coffee?

Augie Dent, Capitola, CA

Caffeine? Yes, please! Americans consume some three hundred million cups of coffee every day. Globally, coffee is second only to oil in terms of dollars traded, and it leaves a whopping social and ecological footprint, particularly in parts of the world with the planet's greatest, and most threatened, biodiversity.

Prior to the 1960s, most coffee was grown under the shady canopies of large trees in conditions similar to tropical forests. These traditional coffee plantations had plant diversity and provided a valuable habitat for migratory birds and other wildlife. All that flora and fauna also helped keep pests in check while providing a wide range of natural nutrients for the soil.

But over the last four decades, feeding the world's coffee addiction began to require greater production, and coffee growers started clearing their land in order to grow higher-yield varieties that thrive in direct sunlight. While financially productive, this sun-grown coffee takes a heavy toll on the environment, on wildlife, and on workers' health by eliminating surrounding biodiversity and requiring heavy use of toxic fertilizers, pesticides, and insecticides.

Starbucks is one coffee-shop chain that's aiming to reverse the trend. In 1998, the company formed a partnership with Conservation International, a leading environmental nonprofit organization, to encourage sustainable, shade-grown coffee production while also ensuring that small farmers and agricultural co-ops earn a living wage for their labors—a practice otherwise known as "fair trade." Starbucks' Organic Shade Grown Mexico and Shade Grown Mexico Decaf coffees are grown with respect for the environment and the economic needs of farmers.

Shade-grown brands are becoming more widely available for home brewers too. The Smithsonian's National Zoo website features a handy listing of "bird friendly" coffee retailers (that is, bean sellers committed to shade-grown coffee only), searchable by zip code. The Rainforest Alliance, which also works to get the word out about coffee's big footprint, certifies several brands, including a line from Yuban and Wal-Mart's Sam's Choice brands. And the website Coffee Review lists Green Mountain, Kaldi's, Thanksgiving Coffee, New Harvest, Kaffe

Magnum Opus, Café Campesino, and Coffee Tea Etc. as topping the list with both excellent taste and environmentally responsible growing practices. Many of these brands are available at organic food specialty stores and at natural foods supermarkets, including Whole Foods and Wild Oats.

CONTACTS: Starbucks, www.starbucks.com; Coffee Review, www.coffeereview.com; Conservation International, www.conservation.org; Rainforest Alliance, www.rainforest-alliance.org; Smithsonian National Zoo "bird friendly" coffee finder, www.nationalzoo.si.edu/conservationandscience/migratorybirds/coffee.

I've enjoyed organic wines, but do any companies make organic beer?

Margaret Chadwick, Weston, CT

If you're going to chug, it's nice to know you can now chug organic. Eighty million Americans regularly drink beer, which breaks down to about twenty-three gallons of the beverage per person every year.

Organic beer is made from organically grown hops, malt, barley, and natural yeast, with no chemical additives or processing. Some other beers are not organic but are brewed along environmental principles, which can meaning anything from 100 percent wind power to hydroponic gardens fed by recycled wastewater. Although no specific sales figures are yet available for organic beer, the category it's in, nondairy beverages, grew 69 percent between 2001 and 2006. In the latter year, the nondairy organics did an incredible $757 million in sales. Although it is more expensive to produce, many organic beers are competitively priced at $5.99 or $6.99 a six-pack.

As the British newspaper the *Telegraph* notes, "Organic beer has to pass the taste test. The finest, greenest credentials count for little if the stuff in the glass is awful." Indeed, many breweries now offer high-quality, tasty products. "I think the future of organic beer is bright," says Crayne Horton, vice president of Fish Brewing Company in Olympia, Washington, which sells Fish Tale Organic Amber Ale in the Northwest. The largest hurdle for organic beer remains achieving national name recognition and distribution, which is the goal of ambitious brewers like California-based Wolaver's, whose organic ales and ciders are now available in thirty-three states.

CONTACTS: Fish Brewing Company, (360) 943-3650, www.fishbrewing.com; Wolaver's, (800) 473-0727, www.wolavers.com.

Is there a way to wash the pesticides off fruits and vegetables, especially those that have been waxed or are hard to clean, before we eat them?

Michelle, Chalmette, LA

Many fruits and vegetables sold in the United States have been treated with pesticides, and residues of these potentially harmful chemicals often remain on their surfaces. Rinsing all produce thoroughly before eating is always a good idea, but many pesticides, fungicides, and other agricultural chemicals are trapped under a wax coating that is often added to resist water and prolong shelf life. In those cases, rinsing produce with just plain water is not enough to do the job. Several companies have developed products that can help.

Organiclean, which contains extracts from coconut, sugarcane, sugar maple, bilberry, orange, and lemon, is completely biodegradable and organic, and is a registered kosher product. Organiclean works well for hard-to-clean produce like strawberries, raspberries, spinach, lettuce, and broccoli and comes in an eight-ounce plastic spray bottle.

Another option is Veggie Wash, from Citrus Magic. Made of natural ingredients from citrus fruit, corn, and coconut, and containing no preservatives, Veggie Wash comes in a sixteen-ounce spray bottle as well as thirty-two-ounce and gallon refills. Fit Fruit & Vegetable Wash is made from citric acid and grapefruit oil and claims to remove 98 percent more pesticides, waxes, and other contaminants when compared to washing with water alone. Fit comes in twelve-ounce spray bottles and thirty-two-ounce refills.

For those inclined to more homespun solutions, various combinations of common pantry items work well too. One recipe calls for soaking produce for five minutes in a fifty-fifty solution of white vinegar and water, while another calls for spraying fruits and vegetables with a combination of one tablespoon of lemon juice, two tablespoons of baking soda, and one cup of water. *Consumer Reports* says that a diluted wash of dish detergent followed by a rinse in tap water eliminates pesticide residues on most fruits and vegetables. After any such treatments, all produce should be rinsed thoroughly in plain water before eating or cooking.

Some analysts say that washing produce is not needed, given strict FDA regulations about pesticide residues. "In the U.S., there's very little produce with pesticide residues anywhere near the allowed tolerance levels," says Elizabeth Andress, a food safety specialist with the University of Georgia's Center for Food Safety. "If you use a produce wash, you may be reducing the levels of pesticide residues," she says, "but the levels were nowhere near harmful to begin with."

The Environmental Protection Agency (EPA) says that the only way to completely avoid pesticide residues is to buy certified organic produce. The majority of supermarkets in the United States now stock pesticide-free organic produce, although it does cost a bit more. Consumers should note, however, that even organic produce should be washed before being eaten, because some pesticide residue is possible (as a result of "pesticide drift" from neighboring crops and other factors), and there may be other impurities caused by human handling.

CONTACTS: EPA, *Pesticides and Food: What You and Your Family Need to Know,* www.epa.gov/pesticides/food; Fit Fruit & Vegetable Wash, (800) FIT-WASH, www.fitwash.com; Veggie Wash, (800) 451-7096, www.citrusmagic.com.

What is the modern meat industry's impact on the environment?

Jeremy Smith, Bellefonte, PA

Just about every aspect of meat production—from grazing-related loss of cropland and open space to the inefficiencies of feeding vast quantities of water and grain to cattle in a hungry world to pollution from "factory farms"—is an environmental disaster with wide and sometimes catastrophic consequences.

A UN study concluded that livestock is a "major player" in climate change, accounting for 18 percent of all greenhouse gas emissions (measured in carbon dioxide equivalents)—more than the entire transportation system!

Meat production has become a major problem because of its very success as a human food. In 1950, world meat production was 44 million pounds annually; today, it has risen fivefold to 253 million tons per year. Pork production, for instance, was less than 5 million tons annually in 1950, but it's more than 90 million tons today. The average person on the planet ate 90.3 pounds of meat in 2003, double the figure for fifty years ago.

The Sierra Club says that producing one pound of grain-fed beef requires about 16 pounds of wheat and—as staggering as it sounds—2,500 gallons of water. Furthermore, millions of acres of forest have been cleared worldwide to make room for the large areas of land needed for cattle grazing. In the United States, more than 260 million acres of forest have been cleared to grow crops to feed animals raised for meat, and an acre of trees disappears every eight seconds.

Tropical rain forests are also being cut to create grazing land for cattle. In some cases, 55 square feet of rain forest may be destroyed to produce just one quarter-pound burger. Since trees absorb carbon dioxide (CO_2), the leading greenhouse gas, this significant loss of forest contributes to global warming as well.

Soil erosion is also mostly due to the meat industry, which, according to the Worldwatch Institute, is directly responsible for 85 percent of all soil erosion in the United States because so much grain is needed to feed the animals. Livestock is fed more than 80 percent of the corn and 95 percent of the oats grown by American farmers. The world's cattle alone consume a quantity of food equal to the caloric needs of 8.7 billion people–more than the entire human population on Earth.

Animal waste is yet another foul outcome of our national meat obsession. It's the largest contributor to pollution in 60 percent of the rivers and streams classified as "impaired" by the EPA. Food animals, according to the EPA, produce waste at a rate of roughly 68,000 pounds per second. Major waste pollutants that make their way into our waterways include nutrients like nitrogen and phosphorus, which can cause massive fish kills; harmful bacteria and viruses; and toxic heavy metals, which are present in some commercial livestock feed.

Critics also point to the fact that meat-based diets exacerbate world hunger. Harvard nutritionist Jean Mayer estimates that a 10 percent reduction in U.S. meat consumption would free up enough grain to feed sixty million people. Some 40 percent of the world's grain harvest is fed to livestock, while nearly a billion people go hungry each day.

While environmental groups recognize the benefits of vegetarianism as an alternative, few recommend it for everyone. Meat-loving environmentalists can look for small farms that feed livestock natural, organic diets; treat animals more humanely; and practice more sustainable land use.

CONTACTS: *E – The Environmental Magazine,* January/February 2002, www.emagazine.com/view/?142; Sierra Club, (415) 977-5500, www.sierraclub.org/factoryfarms; Worldwatch Institute, (202) 452-1999, www.worldwatch.org; the UN report is available free at www.virtualcentre.org/en/library/key_pub/longshad/A0701E00.pdf.

In light of concerns about mercury-tainted fish that have been in the news lately, which fish are safest to eat?

Renee Scott, via e-mail

What goes up must come down. In other words, U.S. power plants spew as much as 150 tons of mercury into the air each year, and eventually that mercury settles into nearby waterways and accumulates in the tissue of fish. Hungry yet? As mercury pollution from industrial facilities spreads in our ocean and freshwater environments, consumers need to limit their fish intake. And that's true whether you catch it yourself or buy it in a supermarket or restaurant.

Nationwide, more than three thousand bodies of water were under fish-consumption advisories in 2003, and the number is climbing. Aquatic predators toward the top of the food chain–including swordfish, shark, king mackerel, tilefish, pike, walleye, largemouth bass, white sucker, yellow perch, and albacore tuna–pack the biggest mercury punch. Environmentalists recommend avoiding eating these fish altogether.

Shrimp, salmon, pollack, catfish, and canned light tuna do not accumulate as much mercury in their systems and are typically safer to eat in moderation. Nevertheless, the EPA recommends that consumers limit their intake to twelve ounces (two average meals) per week of any fish.

Exposure to mercury can be particularly hazardous for pregnant women and small children. During the first several years of life, a child's brain is still developing and rapidly absorbs nutrients. Prenatal and infant mercury exposure can cause mental retardation, cerebral palsy, deafness, and blindness. Even in low doses, mercury may affect a child's development, delay walking and talking, shorten attention span, and cause learning disabilities.

In adults, mercury poisoning can adversely affect fertility and blood pressure regulation and can cause memory loss, tremors, vision loss, and numbness of the fingers and toes. A growing body of evidence suggests that exposure to mercury may also lead to heart disease.

The federal Centers for Disease Control and Prevention says that one in six women of childbearing age have mercury in their blood above the level that would pose a risk to a developing fetus. The good news for consumers who have eaten large amounts of fish in the past is that they can significantly lower the mercury content in their bloodstreams by cutting consumption now.

CONTACTS: EPA "Fish Advisories" page, www.epa.gov/waterscience/fish; Natural Resources Defense Council, "Mercury Contamination in Fish: A Guide to Staying Healthy and Fighting Back," www.nrdc.org/health/effects/mercury/index.asp.

While I love chocolate, I've heard that cocoa bean agriculture is environmentally destructive and exploits workers in tropical rain forests around the world. Is this true?

Dana Lee, Boston, MA

For centuries, small farmers across the world's tropical rain forest zones have been cultivating native strains of cocoa in a sustainable, chemical-free environment. To this day, these small cocoa farms, with their wide variety of plant life and profusion of shade, are known for the plethora of wildlife—including howler monkeys, ocelots, and parrots.

However, as the global appetite for chocolate has increased over the last half century, industrial cocoa farms have taken over the majority of production, clearing vast tracts of tropical rain forest and planting low-quality hybrid cocoa designed to flourish in open fields in the hot sun, with the help of tons of fertilizers and pesticides.

These intensive farming practices have increased erosion and runoff, reduced soil fertility, contaminated crucial water supplies, and destroyed wildlife habitats. And the large-scale nature of the production processes has drawn entire communities into the labor pool, leading to violations of fair labor standards in places where enforcement, let alone the rule of law itself, is often nonexistent. In some countries, including the Ivory Coast, there have been charges of enslavement of cocoa workers.

In response to this unfortunate trend, many cocoa "cooperatives" have sprung up around the world to help small farmers get their products to market while protecting the environment and enabling them to earn a living wage. The New York–based Rainforest Alliance has been instrumental in helping these small cooperatives compete on the world stage.

Since 1999, the group has worked with farm cooperatives to develop guidelines for environmentally sound and socially responsible cocoa production. As a result, more than two thousand small growers now raise cocoa without the use of pesticides and under the shaded canopy of the rain forest, just like their ancestors did for centuries.

In some cases, the very success of such cocoa cooperatives can also hurt the rain forest by encouraging more people to start their own small cocoa farms in and thus displacing the existing tropical rain forest. In the long run, this piecemeal destruction could take a significant toll on the character and biodiversity of the world's tropical rain forests. According to Melissa Schweisguth of the international human-rights organization Global Exchange, much of the problem is economic: "Farmers are cutting the rainforest because they're impoverished and can make money selling timber and growing cocoa." Schweisguth reports that more than half of the rain forest in the Ivory Coast, for instance, has been cut for cocoa plantations.

Global Exchange says that organically grown cocoa sold according to fair trade standards benefits more than forty-two thousand farmers in eight countries. Consumers interested in purchasing organic fair trade cocoa should look for brands certified by TransFair USA, which monitors international fair trade standards on cocoa and a wide variety of other products. U.S. candy makers are clearly getting serious about these issues. For instance, Theo Chocolate in Seattle, Washington, has become the only roaster of organic cocoa beans and the first roaster of certified fair trade cocoa beans in the United States.

CONTACTS: Global Exchange, (415) 255-7296, www.globalexchange.org; Rainforest Alliance, (212) 677-1900, www.rainforest-alliance.org; TransFair USA, (510) 835-0179, www.transfairusa.org.

What does "dolphin-safe tuna" mean, and how can I make sure that the tuna I buy is "dolphin safe"?

Charlie Vestner, San Francisco, CA

Biologists estimate that since the beginning of large-scale commercial fishing in the late 1950s, more than ten million dolphins have been drowned when inadvertently snared in the huge underwater drift nets meant to catch tuna and other fish. Drift nets, which can extend fifty miles as they are left to drag overnight, are indiscriminate killing tools often referred to as "walls of death." In addition to dolphins, large numbers of whales, sharks, and other nontarget species die every day in drift nets. The industry refers to these as "bycatch," and they are usually just tossed back overboard.

Drift-net fishing has been illegal in American waters since the passage of the Marine Mammal Protection Act (MMPA) in 1972. Yet seafood companies were able to source their products from fishing fleets from other countries not subject to U.S. law. By the late 1980s, fishing fleets around the world were deploying some thirty thousand miles of netting daily to meet the growing demand for seafood.

After a public outcry over the needless killing of dolphins, Congress amended the MMPA in 1990 to establish a "dolphin-safe" labeling system so consumers could find tuna caught without the use of drift nets. In 1993, the UN instituted a global moratorium on drift-net fishing. Biologists estimate that these measures have saved millions of dolphins over the last decade.

However, since the early 1990s the U.S. government has gradually been loosening the standards that allow companies to use the "dolphin-safe" label on their cans. In 1995, the World Trade Organization pressured the Clinton administration to lift its embargo on tuna from Mexico and other countries less concerned about the harmful effects of drift-net fishing.

The Bush administration sought to further weaken dolphin-protection efforts by allowing for the importation of drift-net-caught tuna as long as fishermen see no visual evidence of dolphin snaring while harvesting their catches. But in April 2007, a three-judge panel of the United States Court of Appeals for the Ninth Circuit unanimously refused to allow the weakening of the "dolphin-safe" label and condemned the failure of the Bush administration and federal agencies to follow science.

Although the U.S. government's definition of "dolphin safe" may not mean what it used to, the top three American tuna sellers—StarKist, Bumble Bee, and Chicken of the Sea—have vowed to avoid the distribution and sale of tuna from fishing fleets that use drift nets. A number of major chain stores—including A&P, Albertsons, IGA, Kmart, Publix, Safeway, and Wal-Mart—stock only dolphin-safe tuna.

Some restaurant chains—such as Subway, Carl's Jr., Olive Garden, and Red Lobster—serve only dolphin-safe tuna. Tuna consumers who stick to these brands, stores, and restaurants will know their lunch did not needlessly kill dolphins.

CONTACTS: Bumble Bee, www.bumblebee.com/faq; Chicken of the Sea, www.chickenofthesea.com/dolphin_safe.aspx; Defenders of Wildlife, Save-the-Dolphins Campaign, (202) 682-9400, www.defenders.org/wildlife/new/dolphins.html; Earth Island, www.earthisland.org/dolphinSafeTuna/consumer; Earth Trust, (808) 261-5339, www.earthtrust.org; StarKist, (800) 252-1587, www.starkist.com/faqs.html.

I've heard that some foods are now being irradiated. Why is this, and what are the implications for our health and safety?

Emily Worden, Monroe, CT

Food irradiation—used to kill bacteria, parasites, and insects in food and to retard spoilage—is actually not new. Research began early in the twentieth century and picked up in the 1950s as part of the U.S. government's "Atoms for Peace" effort to find non-wartime uses for nuclear technology. The FDA began approving food irradiation in 1963 to rid wheat and flour of insects and to control the sprouting of potatoes. It later approved the irradiation of spices and seasonings to fight insect infestations, and then pork (to prevent trichinosis), poultry (to prevent salmonella and other foodborne bacterial pathogens), and more recently beef, lamb, and pork (to kill *E. coli*).

The World Health Organization calls the practice "sound food-preservation technology."

In recent years, a series of highly publicized events led to increased use of irradiation. In 1998, Sara Lee recalled millions of pounds of hot dogs and deli meat after twenty-one people died in a listeria outbreak.

In 2000, a young Milwaukee girl died after eating watermelon splashed with *E. coli* at a Sizzler restaurant. The *E. coli*, which made six hundred other people sick, was traced to a Colorado meat plant. In 2002, Conagra recalled 19 million pounds of *E. coli*–contaminated beef. There are some thirty-three million cases of food-related illnesses each year, and nine thousand deaths. Food poisoning caused by *E. coli* affects up to twenty thousand people annually.

The FDA says irradiation is safe, but critics charge otherwise. Irradiation does not make food radioactive, but it can create toxic byproducts and some "unique radiolytic products" that haven't yet been identified or tested, says Dr. John W. Gofman of the University of California at Berkeley. "We know that irradiation causes a host of unnatural and sometimes unidentifiable chemicals to be formed within the irradiated foods," he says. "Our ignorance about these compounds makes it simply a fraud to tell the public 'we know' irradiated foods are safe to eat." The Organic Consumers Association (OCA) claims that irradiation saps food's nutritional value and charges that irradiation deactivates raw food's natural digestive enzymes and encourages fats to turn rancid.

Caroline Smith DeWaal, director of food safety at the Center for Science in the Public Interest (CSPI), says that irradiation's benefits outweigh its risks but fears irradiation may be seen as a "silver bullet," leading to neglect of effective sanitation measures in the production of food in the first place. Patty Lovera of the consumer advocacy organization Public Citizen agrees. "People are getting sick because cattle are crowded into small pens, sleeping in their own waste," she says. "Then they move through slaughter so quickly that mistakes cause fecal matter to contaminate the meat." Even the pro-irradiation American Dietetic Association admits, "The process is not a replacement for proper food handling practices."

CONTACTS: American Dietetic Association, (800) 877-1600, www.eatright.org; Center for Science in the Public Interest, (202) 332-9110, www.cspinet.org; Organic Consumers Association, (218) 226-4164, www.organicconsumers.org; Public Citizen, (202) 588-1000, www.citizen.org.

Are raw foods healthier to eat than cooked foods?

Kris Amitzboll, Coledale, Australia

Raw food devotees believe the "living foods" (as they call them) are much healthier than anything cooked or processed. Followers of diets based wholly or largely on raw foods claim numerous health benefits, including increased energy levels, clearer skin, better digestion, weight loss, and reduced risk of heart disease.

A pure raw food diet consists of unprocessed, usually organic, whole-plant-based foods, such as fresh (or dried) fruits and vegetables; nuts and seeds; beans, grains, and legumes; other organic or natural foods that have not been processed; and freshly made fruit and vegetable juices.

The Living and Raw Foods website says that raw, uncooked foods contain essential food enzymes that help the digestion process without relying on the body to produce the enzymes that are lost through cooking. It is also thought that cooking (heating foods above 116 degrees Fahrenheit) destroys vitamins and minerals, and that cooked foods take longer to digest and tend to allow partially digested fats, proteins, and carbohydrates to clog up our digestive system and arteries. In *Living Cuisine: The Art and Spirit of Raw Foods*, raw food chef to the stars Renée Loux Underkoffler argues, "Raw foods make optimal assimilation of nutrition easy, provide pure, clean energy for the body, and do not require a lot of energy for digestion."

Traditional nutrition experts refute this idea, though the American Dietetic Association (ADA) and others are studying the possible benefits of a raw food diet. Claudia Gonzalez, a registered dietitian and spokesperson for the ADA, says that eating all raw, all the time, is an "extreme" diet, but acknowledges, "If you eat more raw foods in your diet [without adding calories] that's always a good thing. Replacing refined, processed foods with raw foods is a healthy move. Eating a few raw meals a week can be great, but it's important not to go to the extreme."

Gonzalez, who has studied raw food diets, says it's hard to eat more than twelve hundred calories a day in raw foods. While this might be great for weight loss, she says, once the weight comes off it might be hard to sustain a person's energy level, especially if they are doing physically demanding work.

If you decide to go raw, there are definitely benefits to the environment. The lower you eat on the food chain, the less impact you will have on the earth's resources. According to Monica Dewart, certified nutritionist and raw foods advocate, "One hundred percent of the 'waste' materials (seeds, peels, etc.) of a raw diet are biodegradable and great for composting. This is the ultimate environmentally friendly diet!"

CONTACTS: Living and Raw Foods, www.living-foods.com; Renée Loux Underkoffler, *Living Cuisine: The Art and Spirit of Raw Foods* (New York: Avery, 2003).

What on earth is this Slow Food movement I keep hearing about?

Robert Davey, Bridgeport, CT

Imagine the eyesore of the McDonald's golden arches perched on the Spanish Steps in Rome. It happened in 1989, to the shock of Italian Carlo Petrini, who channeled his disgust into the international Slow Food movement. Its head offices are in Piedmont, in the north of Italy. More than half of the organization's membership is in Italy, but the group boasts more than seventy-seven thousand members in forty-eight countries, including the United States, which claims seventy-four local chapters. There are currently chapters in Washington, D.C., New York City, Los Angeles, and New Orleans and also in smaller places like Fargo, North Dakota, and Small Green Island, Washington.

The main thrust of Slow Food is to preserve and encourage traditional foods, beverages, and recipes that are "endangered by McNuggets and Monsanto," Petrini says, referring to both our obsession with unhealthy fast food and the increasing and uncertain role of biotechnology. "It's a union of education, politics, environment, and sensual pleasure," says Petrini. The goal: the propagation of leisurely, more epicurean eating habits, and a more enlightened and patient approach to life in general.

"Slow Food is an international movement dedicated to saving the regional cuisines and products of the world," says Patrick Martins, president of Slow Food USA. "It could be style: barbecue, Cajun, creole, organic . . . anything that's fallen by the wayside due to our industrial food culture." Slow Food's primary focus is on saving endangered ways of life that revolve around the stomach. For Slow Food, animals and plants are threatened but so are recipes, harvesting methods, and production techniques.

Slow Food calls its local chapters "convivia." Members organize food and wine events and other initiatives to create "conviviality" and promote the cause. According to Marsha Weiner, who leads the two-hundred-member Washington, D.C., chapter, "Each chapter is very different and independent. Here in D.C. we organize farm visits, hands-on demonstrations with chefs in their kitchens, lectures, and social events."

The sixteen-member State College, Pennsylvania, chapter organizes potluck dinners, lectures, and educational trips. Says coleader Anne Quinncorr, "Mass-produced food had the good intention of getting more affordable food to the greatest number of people. But, there was no foresight given to environmental impact. A peach grown by a small-scale suburban farmer may be a bit more expensive, but it tastes like a peach and when you buy it, you're keeping that farmer in business and fighting urban sprawl."

Slow Food advocates are settling in for a long struggle, but they say victory will eventually be theirs. On the day fast food dies, says Martins, "We will raise a glass of organic wine and say good riddance."

CONTACTS: Slow Food USA, (718) 260-8000, www.slowfoodusa.org; Slow Food (main office), www.slowfood.com.

Do urban gardens significantly contribute to our food supply?

Wayne Chow, New York, NY

Urban gardens, like the ones springing up all over New York City and Seattle, provide 15 percent of the world's food supply. In the United States, they are also creating sorely needed jobs in neglected neighborhoods and introducing inner-city kids to the wonders of nature. Gardens bolster community pride and eliminate some of the environmental problems of modern agribusiness such as heavy use of pesticides and pollution from long-distance transportation.

Town planners often view urban agriculture suspiciously, worried that constituents will be offended by manure and dirt. But the success of city gardens is well founded. Hong Kong, one of the world's most densely populated cities, produces about half of its vegetables in urban gardens. In Moscow, nearly 65 percent of families engage in some kind of food production. In Cuba, according to the Institute for Food and Development Policy (also known as Food First), urban gardens play a crucial role in feeding the country's citizens. Havana, where nearly 20 percent of Cuba's population lives, is home to over eight thousand community gardens, which are cultivated by more than thirty thousand people and cover nearly 30 percent of the available land.

Back in the United States, South Central Los Angeles's Food from the 'Hood program has brought attention to the potential of its embattled Crenshaw district, while providing college funds for the high school students who maintain organic gardens. San Francisco's Fresh Start Farms employs homeless families to grow produce, which is then sold to local restaurants. Even some U.S. prisons have now started urban gardens, which can be on rooftops as well as on the ground.

CONTACTS: Food from the 'Hood, (888) 601-FOOD, www.foodfromthehood.com; Fresh Start Farms, (415) 487-9778, www.grass-roots.org/usa/fresh.shtml; Institute for Food and Development Policy, (510) 654-4400, www.foodfirst.org.

Which is healthier to eat: salmon grown on fish farms or wild salmon?

Jay Simms, Madison, WI

Most of that pink, fatty fish glistening in mounds on a bed of shaved ice at the local supermarket (or grilled at the local restaurant) comes from a salmon farm. Over 50 percent of the world's salmon is now farmed rather than caught in the wild, according to the National Audubon Society.

Farmed salmon is thought to contain ten times the level of PCBs (polychlorinated biphenyls) and other pesticides compared with their wild brethren, probably due to the amount of toxins in their feed. However, a study by the Canadian Food Inspection Agency determined that the levels of these harmful chemicals in fish feed would not cause fish to exceed the safety standards set in the Canadian Guidelines for Chemical Contaminants and Toxins in Fish and Fish Products.

Though there is not yet conclusive evidence about the toxicity of farmed salmon, what is certain is that salmon aquaculture along the Atlantic coast is causing the population of wild Atlantic salmon to decline. According to the National Audubon Society, farms are situated in prime locations where the tide flushes out the area, forcing wild stocks to migrate to less-than-ideal areas. The international North Atlantic Salmon Conservation Organization (NASCO) is working to increase the population of wild salmon stocks through habitat protection and reclamation.

CONTACTS: National Audubon Society, (212) 979-3000, www.audubon.org; North Atlantic Salmon Conservation Organization, (011) 44-131-228-2551, www.nasco.int.

How do I know if what I am buying is genuinely organic?

Michael Faber, Acton, MA

With organic produce and food, it's as simple as checking for the green and white "USDA Organic" sticker. With other items, it's a crapshoot. "The USDA doesn't regulate household cleaners, body care products, and other household items," says Craig Minowa, environmental scientist at the Organic Consumers Association, "and that's where fraudulent labeling occurs."

He says that consumers need to be careful when shopping for organic goods, because companies currently add minimal or questionable organic ingredients to traditional toxics so they can market the product as "all natural" or "100 percent organic."

One such trick involves the use of hydrosol, a by-product of distilling organic plant matter and making essential oils. Companies are capturing the steam created in this process and using this "organic water" as the base of products, allowing them to claim that a large percentage of the product is organic. "The use of hydrosols is a big loophole in organic standards right now," says Minowa. "They're used to inflate organic labeling percentages."

Product labeling has also made headlines lately because of the controversy surrounding genetically modified organisms (GMOs). Congressman Dennis Kucinich (D-OH) sponsored a bill, so far unsuccessful, to require labeling of such food. However, GMO labeling in other countries is taking off. The United States has only a voluntary labeling program created by the National Organic Standards Board. The organic standards prohibit the use of genetically modified organisms, irradiation, sewage sludge, or antibiotics in organic meat and poultry production and require 100 percent organic feed for organic livestock. There are three different labels, based on the percentage of organic ingredients in a product: "100 percent organic"; "organic" (contains at least 95 percent organic ingredients by weight, excluding water and salt); and "made with organic" (contains between 70 and 95 percent organic ingredients). To help consumers, the USDA has designed a seal that can be used on products labeled as "100 percent organic" or "organic."

Organic labels are showing up on many types of products, but just because your T-shirt is made of certified organic cotton doesn't mean that it was treated with chemical-free, organic dyes. Sandra Marquardt, project coordinator for the Organic Trade Association's (OTA) Organic Fiber Council, says that while certification for organic cotton already exists, the OTA is in the final stages of developing organic fiber processing standards that would address the treatment and dyeing of organic cotton. And as for ensuring that the organic T-shirt wasn't stitched in a sweatshop, that's a whole other label battle.

CONTACTS: Organic Consumers Association, (218) 226-4164, www.organicconsumers.org; Organic Trade Association, (413) 774-7511, www.ota.com.

2
THE ENLIGHTENED SHOPAHOLIC

From Wal-Marts to Small Marts

We've got an addiction in this great land of ours. An addiction to stuff. We display buttons and bumper stickers boasting of our consumption habits and drive to warehouse-size stores where we load huge vats of ketchup and extra-thick rolls of toilet paper into giant shopping carts. It's as if a catastrophe lurks around every corner and we are always in the midst of "stocking up" for the inevitable fallout. Well, not to invoke thoughts of fear and calamity (especially when there's a sale on!), but we are facing a disaster of sorts, and one that can be mitigated by our decisions about where we push our shopping carts and what we put inside. Consumer demand has already led to real green changes in company policy at places like Wal-Mart, as well as in production methods. Every ordinary grocery store now carries fair trade coffee and organic produce, and health concerns are leading many to the green side of the aisle. Curbing consumption is always wise, but shopping smart can also accomplish some serious good.

What is the environmental impact of America's buying habits?

Jenni Perez, Seaside, FL

Your next-door neighbor just bought a Hummer. That long-untouched parcel of land around the corner is now home to a new strip mall. And on your short bicycle trip to the office you count dozens of discarded soda cans and bottled water containers with pretty nature scenes on them. Back home, your kid's floor and closet are littered with CDs, video game cartridges, $150 sneakers, and bean-filled toys. Indeed, a recent *Time*/CNN poll found that 80 percent of people think children are more spoiled today than the kids of ten or fifteen years ago. Arguably, the adults may be too.

The United States consumes more energy, water, paper, steel, and meat per capita than any other country, so much so that environmentalists say that at least four additional planets would be needed to provide the American lifestyle to every person on Earth. Meanwhile, forests are being lost at an alarming rate, farmlands and wetlands are being engulfed by development, plant and animal species are disappearing, and our air and water continue to be threatened by pollution.

Participants in the Center for a New American Dream's Web-based

Turn the Tide program follow "nine little actions" to try to reduce their personal impact on the environment, including skipping car trips, eating one less beef meal a week, reducing water use, and installing energy-efficient lightbulbs. The program enables participants to track the positive impact of their actions–and see the cumulative impact of all of the program's participants across North America. CNAD estimates that for every thousand people who pursue the program for one year, 48.5 million gallons of water and 170 trees are saved, and 4 million pounds of carbon dioxide emissions into the atmosphere are avoided.

CONTACT: Center for a New American Dream, (301) 891-3683, www.newdream.org.

How can I find out which companies may be polluting my community?

Mike Butler, Houston, TX

You have a right to know, but don't think the government has always made it easy for you. Polluter information has been publicly available since passage of the Emergency Planning and Community Right-to-Know Act of 1986, but the public was not able to access it easily until the Internet became widely available.

The easiest-to-use source of such information is Scorecard, a free online service provided by the Environmental Defense Fund. Steer your Web browser to the Scorecard Web page, cough up your zip code, and you'll find a Pollution Report Card providing easy-to-read information on polluters and their pollutants in your locale. At the bottom of every Scorecard report are links to help you take action, with options ranging from e-mailing your governor to urge support for tougher air-quality controls to faxing the companies responsible for polluting your air or water.

If Scorecard can't provide the information you need, other options exist. The Right-to-Know Network (RTK Net) provides free access to numerous government and scientific databases that track environmental trends. The service allows users to identify specific factories and their environmental threats and also provides information on the demographics of affected communities.

The EPA has made its data much more accessible to the general public via its own website, where users can find detailed information on specific types of pollutants and their environmental threats. The agency's TRI Explorer, for instance ("TRI" stands for "toxic release inventory"), allows you to search from coast to coast by zip code, state, or county for spills and accidental emissions of toxic chemicals.

For more general information on global environmental trends, the United Nations Environment Programme's (UNEP) GEO Data Portal contains national, regional, and global statistics as well as maps and graphs covering themes such as fresh water, population, forests, emissions, climate, and health trends. The site's snappy technology displays data quickly in several user-friendly formats.

CONTACTS: EPA Toxic Release Inventory Program, (202) 566-0250, www.epa.gov/tri; Right-to-Know Network, (202) 234-8494, www.rtknet.org; Scorecard, www.scorecard.org; United Nations Environment Programme's GEO Data Portal, http://geodata.grid.unep.ch.

What environmental impact should our community expect if we allow Wal-Mart to open up a store nearby?

Sara Jones, Davenport, IA

With more than six thousand stores spread out across the globe, Wal-Mart is the world's biggest retailer, hands down, and also a magnet for criticism for its low wages, inadequate health coverage, and effect on struggling downtowns. Wal-Mart has also had its share of environmental problems.

Environmentalists complain that the company's stores—often on the outskirts of rural communities—eat up open space, replacing farms and forests with concrete and pavement. And the company has been fined repeatedly for environmental negligence. For example, in 2005 Wal-Mart paid $1.15 million in fines to the state of Connecticut for improper storage of pesticides and other toxins that polluted streams near its stores there, according to the website wakeupwalmart.com.

A year earlier, Florida fined the company $765,000 for violating petroleum storage tank laws at its auto service centers. The company admits that it failed to register its fuel tanks and to install devices that prevent overflow, that it did not perform monthly monitoring, and that it blocked state inspections. That same year, Georgia fined Wal-Mart $150,000 for contaminating water outside Atlanta. And the EPA penalized the company $3 million in 2004 for violating the Clean Water Act in nine states. The company was also forced to change its building practices to prevent future water contamination. This came on the heels of a $1 million fine for Clean Water Act violations at seventeen locations in four other states. Wal-Mart also agreed to establish a $4.5 million environmental management plan to improve its compliance with environmental laws at construction sites.

Wal-Mart says that change is afoot within the company. Environmentalist and former Sierra Club president Adam Werbach has been hired as a consultant (a deal with the devil, some say) and "green" Wal-Mart stores have been opened. CEO Lee Scott has said that sustainability in all its forms is a key concern in moving forward. "As one of the largest companies in the world, with an expanding global presence, environmental problems are our problems," Scott told company employees last October.

Scott's green vision includes powering facilities and fleet with renewable energy, cutting back on waste, and selling green products, including compact fluorescent lightbulbs (CFLs). Wal-Mart reportedly crafted its plan with the help of former vice president Al Gore, and commitments include reducing greenhouse gas emissions by 20 percent at existing locations and investing $500 million in environmental improvements each year.

Wal-Mart is also offering produce that is both organic and locally produced. According to Ron McCormick, an executive in the company's produce division, Wal-Mart buys a wide variety of produce based on what's available in each region, instead of shipping produce across the country. "Our whole focus is: How can we reduce food-miles?" he says.

The green attitude also extends to other products, with the company increasing offerings of sustainably harvested fish and organic cotton clothing and bedding. But despite all this, critics are not assuaged; they say Wal-Mart is so focused on profit that it can never truly be green.

CONTACTS: Wake Up Wal-Mart, www.wakeupwalmart.com; Wal-Mart, www.walmartstores.com.

How can ordinary people convince corporations to behave more responsibly toward the environment?

James B., Buckeye, AZ

Beyond the simple exercising of your purchasing power, there are many actions consumers can take—and helpful available resources, some online—to pressure companies to go green.

A good first step is to research corporate environmental records. The websites buyblue.org and alonovo.com evaluate companies according to various "green" criteria. And Co-op America's *Guide to Researching Corporations,* available online, offers everything from corporate product-safety records and animal testing policies to rain forest impact and the air quality in minority neighborhoods.

Co-op America also works at the cutting edge of consumer activism, pushing companies into "doing well by doing good." Its Adopt-A-Supermarket Campaign uses the power of individuals to pressure grocery stores into carrying more fair trade products, including coffee and chocolate made by companies that commit to sustainable environmental practices and guarantee workers fair wages. At Co-op America's website you can download a campaign guide that provides background on the issue and tips on how to form an "adoption team" of concerned citizens that makes regular visits to educate store managers.

Another effort, the Be Safe Campaign, conducted in partnership with the Center for Health, Environment and Justice, encourages major companies to phase out their use of the highly toxic plastic polyvinyl chloride (PVC). They've already convinced Microsoft, Johnson & Johnson, Victoria's Secret, Sears, Kmart, Target, and Bath and Body Works to phase out PVC in their packaging. Other Co-op America successes include persuading Sempra Energy, the parent company of Southern California Gas and San Diego Gas and Electric, to abandon plans to build coal-fired power plants in Nevada and Idaho, and convincing the U.S. Postal Service to withdraw a proposal to deliver all residential mail in blue plastic bags similar to those used for newspapers.

Another group, Ecopledge, recruits consumers to demand specific improvements to companies' environmental behavior. People taking part promise to stop doing business with companies that don't go green. With names in hand, Ecopledge has succeeded in convincing Dell and Apple to reduce the amount of e-waste they generate, getting ConocoPhillips and BP to drop out of Arctic Power (a lobbying entity pushing to open up the Arctic National Wildlife Refuge to oil drilling), and working with Staples and Office Depot to craft green-friendly paper-sourcing policies.

Ecopledge is currently working on a campaign to pressure major rental car companies—including Enterprise, Hertz, Cendant, and Vanguard—to buy and rent cleaner running cars. Rental companies buy 10 percent of all new cars—a total of 1.7 million vehicles per year. If they averaged forty miles per gallon, they would save half a billion gallons of gasoline per year, plus reduce CO_2 emissions by fourteen billion pounds.

Ecopledge is also pressuring major meat producers—including Premium Standard Farms, Smithfield, and Tyson—to clean up hog and other animal waste. Its third campaign is aimed at permanently protecting the Arctic National Wildlife Refuge in Alaska from oil drilling.

CONTACTS: Alonovo, www.alonovo.com; Be Safe Campaign (PVC), www.besafenet.com/pvc; BuyBlue, www.buyblue.org; Co-op America, www.coopamerica.org; Ecopledge, www.ecopledge.com.

I'm bothered by packaging waste—from water in self-serve bottles to all the foil and cardboard you have to break through to get to a new print cartridge. What is being done to make packaging more "green friendly"?

Jeanne L., Canton, CT

Thanks to forward-thinking action by the European Union (EU), people around the world are beginning to recognize that wasteful packaging puts unnecessary stress on the environment. In 1994, the EU issued a directive putting the responsibility of waste reduction and reclamation on manufacturers instead of on retailers, consumers, and local governments.

The program, popularly known as "producer pays" or Extended Producer Responsibility, requires companies to either take back their packaging (by having customers leave it behind in the store or send it in the mail at the producers' expense) or pay a fee to an organization called Green Dot that will handle it for them. Green Dot is now the standard take-back program in two dozen European countries.

Bette Fishbein of the nonprofit INFORM says the concept has "spread like wildfire" and has been adopted by many industrialized nations–including Poland, Hungary, the Czech Republic, Japan, Korea, and Taiwan–but not yet by the United States, which could certainly benefit. According to the EPA, annual generation of municipal solid waste in the United States increased from 88 million tons in 1960 to 229 million tons in 2001, with containers and packaging making up almost a third of the weight.

Some states are interested in the idea, including Maine, which has followed the European model and initiated its own "producer pays" program; the first in the United States. Maine requires electronics makers to fund consolidation centers where used TV and computer monitors are sent. Maine's Department of Environmental Protection boasts that its law "is a national model, as it protects our environment, saves taxpayers money, and puts costs where they belong to encourage safe design and recycling of electronic wastes."

Some U.S. companies are also taking initiative. Microsoft worked with Packaging 2.0, a packaging-solutions company that recycles used materials into new packaging, to develop an environmentally responsible and reusable package for its line of GPS consumer electronics products. And a number of other companies–including Unilever, Johnson & Johnson, Kraft Foods, and Nike–have come together under the umbrella of the Sustainable Packaging Coalition, a project of the nonprofit GreenBlue, and released a guide for designers and developers to assist them in designing sustainable packaging.

In 2008, Wal-Mart began using a "packaging scorecard" to measure and evaluate its entire supply chain. Goals include using less packaging and using more sustainable materials in packaging. According to Wal-Mart, the company is already beginning to make headway. "By reducing the packaging on one of our patio sets," says the company website, "we were able to use 400 fewer shipping containers to deliver them. We created less trash, and saved our customers a bundle while doing it."

CONTACTS: Green Dot, www.packaging-waste.com; INFORM, www.informinc.org; Sustainable Packaging Coalition, www.sustainablepackaging.org.

What is the deal with plastic recycling these days? Can you explain what the different numbers molded onto the bottom of plastic containers stand for?

Tom Croarkin, Cleveland, OH

Those little recycling triangles should be as easy as one, two, three, but the reality is much more complicated. Plastics, in their infinite variety, are especially troublesome, requiring different processing in order to be reformulated and reused as raw material. Some municipalities accept all types of plastic for recycling, while others only accept jugs, containers, and bottles with the right numbers stamped on their bottoms.

The symbol code we're familiar with—a single digit ranging from 1 to 7 surrounded by a triangle of arrows—was designed by the Society of the Plastics Industry (SPI) in 1988 to allow consumers and recyclers to differentiate types of plastics while providing a uniform coding system for manufacturers.

The numbers, which thirty-nine states now require be molded or imprinted on all eight-ounce to five-gallon containers that can accept the symbol (minimum size, a half inch), identify the type of plastic and, according to the American Plastics Council, an industry trade group, help recyclers do their jobs more effectively.

The easiest and most common plastics to recycle are made of polyethylene terephthalate (PET) and are assigned a 1. Examples include soda and water bottles, medicine containers, and many other common consumer product containers. Once it has been processed by a recycling facility, PET can become fiberfill for winter coats, sleeping bags, and life jackets. It can also be used to make bean bags, rope, car bumpers, tennis ball felt, combs, cassette tapes, sails for boats, furniture—and, of course, other bottles.

Number 2 is reserved for high-density polyethylene plastics. These include those heavier containers that hold laundry detergents and bleaches as well as milk, shampoo, and motor oil. Number 2 plastic is often recycled into toys, piping, plastic lumber, and rope. Like PET plastic, it is widely accepted at recycling centers.

Polyvinyl chloride, commonly used in plastic pipes, shower curtains, medical tubing, vinyl dashboards—even some baby bottle nipples—gets number 3. Like numbers 4 (wrapping films, grocery and sandwich bags and other containers made of low-density polyethylene) and 5 (polypropylene containers used in Tupperware, among other products), few municipal recycling centers will accept it due to its very low rate of recyclability. Number 6 goes on polystyrene (Styrofoam) items such as coffee cups, disposable cutlery, meat trays, packing "peanuts," and insulation and is widely accepted because it can be reprocessed into many items, including cassette tapes and rigid foam insulation.

Finally, anything made from combinations of plastics or from unique plastic formulations not commonly used get imprinted with a 7 or nothing at all. These plastics are the most difficult to recycle and are seldom collected or recycled. More ambitious consumers can feel free to return them to the product manufacturers to avoid contributing to the local waste stream and instead put the burden on the makers to recycle or dispose of the items properly.

CONTACTS: American Plastics Council, www.americanplasticscouncil.org; Society of the Plastics Industry (SPI), www.socplas.org.

Where can I find green-friendly office products and back-to-school supplies?

Taylor Howe, San Diego, CA

A decade ago, the category barely existed outside of specialty stores, but now many recycled papers, pens, pencils, ink toner cartridges, binders, folders, and desk accessories have made it to the big office chains.

Paper use continues to be the largest source of waste generated by office workers and students, and several paper manufacturers have risen to the challenge of providing recycled and even "tree-free" papers at competitive prices. New Leaf Everest, Badger Envirographic, and Eureka 100 are some of the leaders in recycled paper. Dolphin Blue makes tree-free paper from recycled scraps of denim, old money, and the plants hemp and kenaf. Buyers can order these papers from online vendors, including GreenLine Paper and Treecycle, but office stores now carry a wide array of 100 percent recycled postconsumer waste papers.

Meanwhile, materials such as biodegradable cornstarch and recycled plastic and cardboard are starting to replace virgin plastic and vinyl in pens, binders, notebooks, and desk accessories like rulers, pencil cases, and staplers. Also, pencil manufacturers such as Pentel, Autopoint, and ForestChoice have gotten serious about crafting their products from sustainably harvested timber and other green materials, including old currency. Online vendors like Green Earth Office Supply, the Recycled Office Products Co., Real Earth Environmental Company, and Mama's Earth stock these products. Meanwhile, Discount Inkjet Printer Ink Cartridges sells a wide range of recycled inkjet toner cartridges and ink refills compatible with all major brands of copiers and computer printers.

Many of these companies offer special price breaks for nonprofits, local government agencies, schools, and universities and donate a portion of proceeds to environmental nonprofits. Consumers shopping at these stores can rest assured that they are minimizing their impact on the earth while supporting small, innovative companies. But those in need of a quick green fix might be surprised at how good the selection is these days at places like Office Depot, Staples, and OfficeMax.

Buying only environmentally friendly office and school supplies is a great place to start in getting out from under the nonsustainable virgin paper chase.

CONTACTS: Discount Inkjet Printer Ink Cartridges, www.discount-inks.com; Green Earth Office Supply, www.greenearthofficesupply.com; GreenLine Paper, www.greenlinepaper.com; Mama's Earth, www.mamasearth.com; Recycled Office Products Co., www.recycledofficeproducts.com; Real Earth, www.treeco.com; Treecycle, www.treecycle.com.

Bottled water companies would have us all believe that tap water is unsafe to drink. But I've heard that most tap water is actually pretty safe. Is this true?

Sam Tsiryulnikov, Los Angeles, CA

Tap water is hardly blemish free. The Environmental Working Group (EWG) in 2005 tested municipal water in forty-two states and detected some 260 contaminants in public water supplies, 140 of which were unregulated chemicals; that is, chemicals for which public health officials have no safety standards, much less methods for removing them.

EWG did find more than 90 percent compliance on the part of water utilities in applying and enforcing standards that exist, but faults the EPA for failing to establish standards for so many of the contaminants—from industry, agriculture, and urban runoff—that do end up in our water.

Despite these seemingly alarming stats, the Natural Resources Defense Council (NRDC), which has also conducted extensive tests on municipal as well as bottled water, says: "In the short term, if you are an adult with no special health conditions, and you are not pregnant, then you can drink most cities' tap water without having to worry." This is because the contaminants in public water supplies exist in only very small concentrations.

As for bottled water, it's important to know that 25 to 30 percent of it comes straight from municipal tap water systems, despite the pretty nature scenes on the bottles that imply otherwise. Some of that water goes through additional filtering, but some does not.

NRDC has researched bottled water extensively and has found that it is "subject to less rigorous testing and purity standards than those which apply to city tap water." Bottled water is required to be tested less frequently than tap water for bacteria and chemical contaminants, and FDA bottled water rules allow for some contamination by *E. coli* or fecal coliform, contrary to EPA tap water rules, which prohibit any such contamination. There are no requirements for bottled water to be disinfected or tested for parasites such as cryptosporidium or giardia, unlike more stringent EPA rules regulating tap water.

The good news is that consumers are starting to turn away from bottled water. There was public outrage in Britain a few years ago when it was discovered that Coca-Cola's Dasani brand, marketed as "pure, still water" and sold for one pound ($1.90) for a half liter, was simply tap water from a public water supply southeast of London. To make matters worse, shortly thereafter the beverage giant had to hastily withdraw half a million bottles when it was learned they contained nearly twice the legal amounts of a carcinogenic chemical (added by Coke during treatment).

Despite the facts, bottled water enjoys a "cool" factor that tap water can never match. A World Wildlife Fund study confirmed that consumers widely associate bottled water with social status and healthy living. But in test after test, most people can't tell the difference between bottled water and tap water. When *Good Morning America* conducted a blind taste test with its studio audience, New York City tap water was chosen as the heavy favorite over Poland Spring, Evian, and the oxygenated water Life O2.

The bottom line is that we have invested considerably in highly efficient municipal water delivery systems that bring this precious liquid straight to our kitchen faucets anytime we need it. Our focus should be on making sure our tap water is clean and safe for all.

CONTACTS: Environmental Working Group, www.ewg.org/tapwater/findings.php; EPA, "Local Drinking Water Information," www.epa.gov/safewater/dwinfo.htm; National Resources Defense Council, www.nrdc.org/water; Turn to Tap, www.turntotap.com.

Why do some people complain about fluoride in drinking water and toothpaste?

Becky Johnston, Shoreline, WA

Communities began adding fluoride to water supplies in the early 1940s after decades of studies into why some Colorado residents were exhibiting a discoloration or "mottling" of the teeth but at the same time very low rates of actual decay. The culprit turned out to be high concentrations of a naturally occurring fluoride that was running off into the water from Pikes Peak after rainfalls. Research later concluded that adding small, controlled amounts of fluoride to public water supplies would act as a form of community-wide cavity prevention without causing the undesirable mottling known at the time as "Colorado stain."

Today, supporters of fluoridation cite federal research from the

Centers for Disease Control and Prevention showing that the very inexpensive fluoridation of drinking water has since correlated to significant reductions in incidences of tooth decay (15 to 40 percent) in communities across the country. But skeptics worry that we may be getting too much of a good thing. While small amounts of fluoride will prevent tooth decay, excessive amounts can lead not only to irreversible tooth discoloration (today called fluorosis) but also to other health issues, including an increased risk of bone breakage and osteoporosis.

The problem, says the Fluoride Action Network (FAN), which is opposed to fluoridation, is that water supplies that are treated for dental purposes are also used in the making of many common food products—from baby formula and cereal to juices, sodas, wines, beers, and even to water fresh produce. And with most toothpastes also adding fluoride, many people are ingesting far more fluoride than they should.

The main concern for most people is the discoloration of children's second teeth once the baby teeth are gone. Besides being embarrassing, there is no cure. And some doctors worry that excessive fluoride may actually be promoting tooth decay rather than preventing it—and harming kids in other ways, particularly as they get older. FAN cites studies showing how low-to-moderate doses of fluoride can lead to eczema, reduced thyroid activity, hyperactivity, IQ deficits, premature puberty, and even bone cancer.

Concerns have also arisen that by drinking nonfluoridated bottled water instead of tap water we may be increasing incidences of tooth decay (though some bottled waters have added fluoride). John W. Stamm, dean of the School of Dentistry at the University of North Carolina and a spokesperson for the American Dental Association, says, "The simple fact that people may be consuming variable amounts of bottled water seems to me to be insufficient reason to be concerned about a fluoride-deficient diet."

Avoiding fluoride is difficult for people whose local water is fluoridated. And the only filters that can strain fluoride out of water are expensive and use reverse osmosis, activated alumina, or distillation. Switching to unfluoridated toothpaste—many varieties are available from natural health retailers—is one way to cut down on fluoride intake, especially for those who swallow toothpaste when they are brushing.

CONTACTS: Centers for Disease Control and Prevention, www.cdc.gov/oralhealth; Fluoride Action Network, www.fluoridealert.org.

What tax or other government incentives are out there for buying green—for individuals as well as for businesses?

Sarah Rafferty, New York, NY

There has never been a better time than now to tap into green tax rebates and other financial incentives. At the federal level, you can reap the rewards of no less than eight different financial incentives, ranging from tax credits and home loans for replacing windows and installing insulation around the house to tax rebates for purchasing a hybrid car or hooking up a solar hot-water heater.

Nearly every U.S. state has additional state or local incentives available. Washington State, for example, charges no sales tax on renewable energy equipment produced or sold there. And some forward-thinking cities are beginning to offer "density bonuses" and green building incentives to developers and builders to encourage sustainable land use.

Many states require utilities to rebate consumers who save electricity. Some utilities even offer "net metering," whereby consumers who generate some of their power through rooftop solar panels or other technologies can sell electricity back to the utility, thus reducing or zeroing out their electric bill—even earning money.

Many financial incentives are in place for businesses as well. At the federal level, examples include an energy-efficient commercial buildings tax deduction, a business energy reduction tax credit, an energy-efficient appliance tax credit for manufacturers, and a new energy-efficient tax credit for green-savvy builders.

The best place to look for what's available is the free online Database of State Incentives for Renewables and Efficiency (DSIRE), a comprehensive source of information on state, local, utility, and federal incentives that promote renewable energy and energy efficiency.

The Office of Energy Efficiency at Natural Resources Canada offers a slate of grants and incentives under its ecoENERGY Retrofit program to homeowners, businesses, large industries, and public institutions, to help them invest in energy- and pollution-saving upgrades. The agency also administers the High Efficiency Home Heating System Cost Relief program, which will contribute up to three hundred Canadian dollars to home owners who upgrade their old oil or gas furnace or boiler to a new high-efficiency model. And low-income households might qualify for additional federal financial assistance for energy retrofits. Another Canadian program, the Vehicle Efficiency Incentive (VEI) rewards people who buy fuel-efficient cars or trucks with rebates of up to two thousand Canadian dollars each.

CONTACTS: Database of State Incentives for Renewable Energy, www.dsireusa.org; Natural Resources Canada's ecoENERGY Retrofit program, www.oee.nrcan.gc.ca/corporate/incentives.cfm.

A number of products, including paper and clothing—even food and beer—are made from hemp. What is it about hemp that makes it so versatile, and why is it illegal to grow in the United States? Is it also illegal in Canada?

Doug Jones, via e-mail

What do the first Gutenberg bible, Christopher Columbus's ropes and sails, the Declaration of Independence, and the first American flag have in common? All were made from hemp. Many of America's forefathers, including George Washington and Thomas Jefferson, once earned a living growing and selling hemp, which was used to make everything from paper and rope to sails and clothing. During World War II, the crop was of such strategic importance for making clothing that the U.S. government provided farmers with subsidies to convert fields to hemp cultivation.

Hemp is a renewable and easy-to-grow crop that is tough enough to substitute for paper or wood and malleable enough to be made into clothing and even a biodegradable form of plastic. Meanwhile, hemp oil is all the rage among natural food gourmands, who enjoy its nutty flavor and its healthy amounts of protein and omega fatty acids. Hemp is also a popular ingredient in many new hand and body lotions.

Environmentalists and farmers alike appreciate hemp as an alternative to cotton for clothes and trees for paper. Unlike cotton, hemp does not require large doses of pesticides and herbicides, as it is naturally resistant to pests and grows fast, crowding out weeds. To make paper, trees must grow for many years, but a field of hemp can be harvested in a few months and over a few decades will make four times as much paper as a similarly sized forest. Also, making paper from hemp uses only a fraction of the chemicals required to turn trees into paper.

In spite of hemp's versatility, in 1970 Congress designated hemp, along with its relative marijuana, a "Schedule 1" drug under the Controlled Substances Act, making it illegal to grow without a license from the Drug Enforcement Administration (DEA). Although industrial hemp does not contain enough psychoactive ingredients to make a smoker "high," farmers who grow it can risk jail time. Today, the United States is the only developed country that has not established hemp as an agricultural crop, according to the Congressional Research Service. Britain lifted a similar ban in 1993, and Germany and

Canada followed suit soon after. The European Union has subsidized hemp production since the 1990s.

With their American competition out of the running, Canadian farmers have been reaping hemp's financial rewards, especially following a ruling by a U.S. federal court that products made from hemp could be imported into the country. In 2005, the Canadian hemp industry tripled the amount of acreage dedicated to the crop to meet the rising demand, according to the Canadian Hemp Trade Alliance.

American farmers are intensifying their lobbying efforts to lift the U.S. ban. State legislatures in Hawaii, Kentucky, Maine, Montana, North Dakota, Arkansas, Maryland, Minnesota, New Mexico, Vermont, Virginia, and West Virginia, have all passed laws either authorizing hemp research or attempting to make it legal if the U.S. government were to allow it.

But a hemp farming bill introduced into Congress this past year by 2008 presidential candidate Ron Paul (R-TX) stalled out due to opposition from the DEA and the White House. For its part, the DEA maintains that allowing American farmers to grow hemp would undermine the "war on drugs," as marijuana growers could camouflage their illicit operations with similar-looking hemp plants.

CONTACTS: Canadian Hemp Trade Alliance, www.hemptrade.ca; Vote Hemp, www.votehemp.com.

What is better for the environment, cork or plastic wine stoppers or screw tops?

Susan Wolniakowski, Duluth, MN

Though you might be surprised, natural cork wine stoppers are the best choice, primarily because harvesting the real stuff is an age-old practice that keeps alive the world's relatively small population of cork oak trees, which can live for hundreds of years. These scattered pockets of cork oaks, mostly in Portugal and Spain, thrive in the hot, arid conditions of the southern Mediterranean, sheltering a wide array of biodiversity and helping to protect the soil from drying out.

Cork oak forests are important for the survival of some important wildlife, including the Iberian lynx and the Barbary deer, as well as rare birds such as the Iberian imperial eagle, the black stork, and the Egyptian mongoose. As wine producers switch to other types of wine stoppers, the cork oak forests could be abandoned and the trees and the myriad plants and animals that depend on them could die out.

Seventy percent of wine bottles still use natural cork stoppers, but plastic and glass alternatives have been gaining ground in recent years. Winemakers around the world are switching to alternatives, citing benefits including freedom from the cork mold that can taint wine and the ability to easily restopper opened bottles. In Australia and New Zealand—both promising players on the global wine scene—the majority of wine producers use screw caps, mainly because they can make them cheaply instead of paying the relatively high price of importing natural cork.

The growing popularity of screw caps and plastic stoppers has cork producers and environmentalists worried. In a recent report, the World Wildlife Fund (WWF) predicts that, at the current rate of adoption by wine producers, screw caps and other synthetic noncork wine stoppers will dominate the market by 2015, calling into question the future of Mediterranean cork forests. WWF is supporting efforts by Portuguese cork producers to certify their practices as sustainable through the nonprofit Forest Stewardship Council (FSC), which promotes sustainable, economically viable forestry practices around the world.

"Cork oak forests rank among the top biodiversity hotspots in the Mediterranean and in Europe," says WWF's Nora Berrahmouni. At the same time, they are the backbone of an entire economy. FSC certification will reinforce the already environmentally friendly characteristics of the cork economy, leading to new opportunities in cork markets."

Public opinion will undoubtedly play a big role. Producers of plastic stoppers and metal screw caps are working hard to overcome stigmas, since most consumers still associate noncork stoppers with cheap wine. Partly for that reason, and partly because the cork reserves are in their own backyards, the world's premiere European winemakers in Europe are still bullish on cork. Wine enthusiasts everywhere can make an environmental choice by choosing bottles with natural cork stoppers.

CONTACTS: World Wildlife Fund report *Cork Screwed?*, http://assets.panda.org/downloads/cork_rev12_print.pdf; Forest Stewardship Council, www.fsc.org.

Are there any environmentally friendly alternatives to using chemical weed killers like Roundup?

Wyatt Walley, Needham, MA

Glyphosate, the active ingredient in Monsanto's Roundup, is a known toxin. No surprise there—the presence of the powerful chemical helps explain why Roundup is so successful at killing pesky weeds. In fact, glyphosate is the most commonly used pesticide in the United States. The EPA estimates that more than five million pounds of it are used annually on American yards and gardens.

According to Caroline Cox, staff scientist at the Northwest Coalition for Alternatives to Pesticides (NCAP), gardeners wouldn't use Roundup if they knew about all of the problems attributed to it. For instance, ingesting about three-fourths of a cup can be lethal. And symptoms of casual contact can include eye and skin irritation, lung congestion, and erosion of the intestinal tract. Monsanto's Roundup has also been linked to cancer, miscarriages, and genetic damage in humans, so it's no wonder that NCAP and other organizations are pushing for safer alternatives. Environmentally, the product is thought to be implicated in immune system damage in fish and reproductive disruption in amphibians.

Over a recent eight-year study period in California, exposure to glyphosate was the third most frequently reported cause of illness related to agricultural pesticide use. And scientists from the National Cancer Institute and three prominent medical centers have shown the use of glyphosate herbicides by midwestern farmers to be associated with many cases of non-Hodgkin's lymphoma. Roundup also contains other nonactive ingredients, contact with which can cause nausea, diarrhea, chemical pneumonia, laryngitis, and severe headaches.

For its part, Monsanto says that its glyphosate-based herbicides "certainly have one of the most extensive worldwide human health, occupational safety, and environmental databases ever completed on a pesticide product." When used as directed, it says Roundup Pro offers effective weed control "without unreasonable adverse effects on the environment or human health."

There are effective pesticide-free solutions to the weed problems in our yards and gardens. For instance, mulches made from wood chips, straw, grass clippings, or shredded bark can be used to keep weed seeds from germinating. By keeping light from reaching weeds, a thick mulch layer naturally inhibits the growth of the chlorophyll that is the lifeblood of fast-growing weeds.

Maintaining healthy, well-aerated soil is essential to a program of chemical-free weed control, and organic fertilizer works really well for that purpose. Allowing your grass to grow until it's between two and three inches tall also helps keep weeds in check without chemicals. For those hardy survivors, nonchemical assistants include hoes, string trimmers, weed pullers, flame weeders, and radiant heat weeders. Local organic nurseries can help you determine which techniques will work best in your area.

One added benefit of giving up the Roundup habit might be the blossoming of beneficial plants, fungi, and creepy crawlies in your yard. Since Roundup is toxic to a wide range of such important ecological builders as ladybugs, beetles, earthworms, and fungi, going without can help bring these species back to work aerating your soil and keeping virulent pests in check naturally.

CONTACT: Northwest Coalition for Alternatives to Pesticides's Healthier Homes and Gardens Program, www.pesticide.org/HHG.html.

Are there sources for disposable cups, plates, napkins, and dinnerware that are more eco-friendly than others?

Charles Phillips, New York, NY

Most of us grew up with disposable dishware as part of our modern "on the go" culture, so we throw it away without a second thought. And that's how nearly one hundred billion plastic, paper, and Styrofoam cups end up in American landfills and incinerators every year. Human health is the real loser when it comes to our consumption of such products, which are typically made from petroleum-based plastics, hazardous foam, or chlorine-bleached virgin paper.

For the eco-conscious host who enjoys entertaining large groups but doesn't want to wash dishes, compostable dishware provides an answer. California-based Sinless Buying makes a wide range of compostable dinnerware—from dishes and cups to cafeteria-style trays and soup bowls—out of bagasse, fully biodegradable organic sugarcane fiber that's a by-product of making sugar. Once this biodegradable dinnerware has served its purpose, it can simply be tossed in with the backyard or garden compost. Sinless Buying also offers unbleached versions of some of its products.

Another California company, Cereplast, makes its highly regarded

NAT-UR line of compostable cups, plates, utensils, straws, and even trash bags out of a plasticlike substance made from biodegradable corn by-products, also completely biodegradable and compostable.

Montana's Treecycle, best known for its wide variety of recycled papers, now also manufactures biodegradable plates, cups, bowls, and trays made from sugarcane by-products, as well as disposable cutlery made from 100 percent compostable wheat wastes. All Treecycle dinnerware can be machine washed and reused several times before composting.

A new retail entrant is EarthShell, which makes a wide range of compostable plates, cups, cutlery, and food storage containers from such renewable materials as limestone and starch. The company has been supplying food service giants like SYSCO and McDonald's for years and now sells direct to consumers under the ReNewable Products brand name, available at Smart & Final stores in the West and at Schnucks in the Midwest.

There are issues with compostable cutlery, however. While some compostables can go into a home compost pile, most cannot. And the country's few commercial composting facilities are clustered in a handful of states. With consumers left holding the bag, the dump beckons. And even bioplastics don't readily disintegrate in landfills.

Martha Leflar of the Sustainable Packaging Coalition in Charlottesville, Virginia, says bioplastics may be premature. "There's no infrastructure yet for collection," she says. The biopolymer industry "is designing something for a system that doesn't exist."

For Leflar, here's the bottom line: "If you plan to compost the food waste and plasticware, then biopolymers are the correct choice. If you're going to recycle the plastic, then purchasing a product with recycling infrastructure available is the correct choice."

CONTACTS: EarthShell, www.earthshell.com; NAT-UR Store, www.nat-urstore.com; Sinless Buying, www.sinlessbuying.com; Treecycle, www.treecycle.com.

I've heard I should avoid buying wood products made from "old-growth timber." What does that refer to, and how can I tell if something is made from old-growth wood?

Anna Hunt, Sierra Madre, CA

The tall, majestic trees that greeted the first Western visitors to America were "old growth," which simply means they'd been growing for two hundred years or longer. The problem, according to the Rainforest Action Network (RAN), is that the lumber industry classifies trees by lumber grades, not age, and because old-growth wood provides the highest quality lumber, it is highly prized. The old growth left in this country (just 5 percent of what once existed) is mostly found in the Pacific Northwest and California.

While there hasn't been much successful legislation to protect old growth, it is possible to trace where your wood comes from and protect old-growth forests by boycotting products made from it, says Richard Donovan, chief of forestry at the Rainforest Alliance, which created the SmartWood forest certification program. "You can identify suppliers and then look at their forest management," he says.

Donovan recommends buying forest products certified by the Forest Stewardship Council (FSC) as sustainably harvested from a well-managed forest and warns that the new certification label from the American Forest and Paper Association, created in 2002 and called the Sustainable Forestry Initiative (SFI) label, is not sufficient.

Few groups outside of the timber industry recognize the legitimacy of the SFI label. FSC-certified wood is favored by the influential U.S. Green Building Council, for instance. Corporate leaders in sustainable wood, including IKEA, Home Depot, and Kinko's, use FSC-certified products in some cases because of pressure from rain forest activists. The Rainforest Action Network says the SFI label fails to protect old-growth forests, roadless areas and federal lands, endangered species, and indigenous rights. RAN also recommends using timber alternatives when possible, such as recycled wood, composite wood made from plastic, and kenaf paper.

CONTACT: Forest Stewardship Council, (877) 372-5646, www.fscus.org; Rainforest Action Network, (415) 398-4404, www.ran.org; Smart Wood, (802) 434-5491, www.smartwood.org.

Which pet foods are the healthiest and most earth friendly? Can I feed my dogs and cats vegetarian?

Carolyn Cacciotti, Charlottesville, VA

With consumer demand for organic food growing rapidly, it's no wonder that pet owners are also starting to think about what they are feeding to Fido and Mittens. And, oh boy, what our cats and dogs are eating!

Warning: gross stuff ahead. Listing the protein source on a pet food can as "meat or poultry by-products" allows the manufacturer to include meat processing waste. This includes "4-D" animals: dead, diseased, dying, or disabled, whose meat often contains tumors and drugs used to try to treat the animals before they died. According to the Association of American Feed Control Officials (AAFCO), pet food can and does include spray-dried animal blood, hydrolyzed hair, dehydrated garbage, unborn carcasses, and many other things.

A few brands stand out for their commitment to doing away with all that and making a firm commitment to all-natural ingredients. Breeder's Choice offers several all-natural lines of age-appropriate dog and cat foods. Other reputable producers, many with organic or hormone-free ingredients, include Iams, Natural Balance, Honest Kitchen, Natura, Yarrah, Eagle Pack, Wysong, and Urban Carnivore, among many others.

Actor Paul Newman's company, Newman's Own Organics, known for providing people with organic salad dressings, pasta sauces, and popcorn (and donating all profits to charity), has a line of healthy pet foods with all profits going to support animal welfare causes. All of the company's pet food varieties contain certified organic ingredients and avoid antibiotics, hormones, chemical ingredients, artificial preservatives, colors, and additives.

Even though your values may dictate vegetarian pets, it's actually quite a challenge. Dogs are omnivores and can survive on a varied diet, but many veterinarians say they do best when at least some animal protein is on the menu. Cats don't have the luxury of choice; they require certain nutrients from meat that they can't get from plant-based foods. Deficiencies can lead to blindness and even death.

Luckily, dog and cat owners whose vegetarian beliefs extend to their pets' diets do have some options. Yarrah's Organic Vegetarian Dog Food, for instance, contains whole wheat, soy, sunflower seeds, maize, yeast, sesame chips, and minerals and is recommended for overweight dogs. Meanwhile, Evolution Diet, available from petfoodshop.com, makes a wide range of healthy vegetarian dog and cat foods that contain nutritional supplements to keep otherwise carnivorous pets healthy without the meat.

The Animal Protection Institute offers a handy online set of guidelines for choosing healthy foods for your dog or cat. While these pet foods can be found at pet stores across the country, they are also starting to appear in health food stores like Whole Foods and Wild Oats.

CONTACTS: Breeder's Choice, www.breeders-choice.com; Eagle Pack, www.eaglepack.com; Evolution Diet, www.petfoodshop.com; Newman's Own, www.newmansownorganics.com/pet; Urban Carnivore, www.urbancarnivore.com; Yarrah, www.yarrah.com.

3

SAY "AAAAH!"

Healthier Living from the Refrigerator to the Medicine Cabinet

When we really start to look at the labels on our bathroom products, we find a whole lot of confusing messages. The front of the package proclaims "all natural" but the complex chemical names on the back sound suspiciously unnatural. And yet these are the products that cover our skin every day: the shampoos, deodorants, toothpastes, and makeup that keep us looking (and smelling) our best. As research uncovers the real chemical makeup of these products, we have to take a closer look at our beauty regimens. How much is our dyed hair or sun-kissed skin really worth? Is there such a thing as healthier water? Do we really have to give up nail polish? Once we're armed with the knowledge of what goes into our products, we can find the truly natural and not just claiming-to-be-natural alternative to the chemically intensive standbys. And all those chemicals—from headache medicines to shampoo residues—end up in our water supplies and leach from landfills. What's good for our health, as is turns out, is good for the planet's health too.

Is there a connection between environmental toxins and breast cancer?

Ben Ward, Virginia Beach, VA

It's strongly suggested. In the United States, more than two hundred thousand women are diagnosed with breast cancer each year, and 20 percent are likely to die from it. Breast cancers among women have climbed steadily in the United States and other industrialized nations since the 1940s. More than half of women diagnosed with breast cancer do not have any of the known or traditional risk factors such as family history, hormonal factors, or a fatty diet, and researchers suspect that widespread exposure to environmental toxins is triggering the surge.

Evidence linking chemicals to breast cancer includes studies showing that lifetime exposure to naturally produced estrogens (female hormones produced by the ovaries and other adrenal glands) increases the risk of breast cancer. New evidence also suggests that exposure to compounds that mimic these natural estrogens, such as hormone replacement therapy and oral contraceptives, also increases risk.

Other compounds found to increase breast cancer risk include polyvinyl chloride, a plastic commonly used in vinyl siding, shower curtains, and other products; the gasoline component benzene; and some pesticides and herbicides. Also strongly linked are organic solvents used in manufacturing processes, hydrocarbons produced by the combustion of gasoline and heating oil, and synthetic chemicals like dioxin, a by-product of the paper bleaching process. Many compounds long ago phased out of use in the United States–including DES, a drug taken by pregnant women to prevent miscarriage, the notorious pesticide DDT, and PCBs used in manufacturing–still persist in the environment and can also trigger the disease.

When New York health researchers noticed that breast cancer cases were increasing at alarming rates on Long Island during the 1980s and 1990s, they commissioned a study to find out if exposure to some prevalent toxins–including DDT and PCBs–was to blame. Surprisingly, researchers found little evidence to support a definitive connection. However, research did suggest that these chemicals were linked to enlarged tumor size, meaning that although they may not cause breast cancer, they may contribute to how fast the cancer grows.

Without many direct links between breast cancer and specific contaminants, regulation is unlikely, so women should take precautions. Exercising more, increasing consumption of fruits and vegetables, lowering alcohol intake, and quitting smoking are good first steps.

Avoiding exposure to contaminants at home or on the job will also help. The Breast Cancer Fund and Breast Cancer Action advocate for more FDA regulation of chemicals and pressing chemical makers to voluntarily limit production of some substances.

CONTACTS: Breast Cancer Action, (415) 243-9301, www.bcaction.org; Breast Cancer Fund's Long Island Breast Cancer Study Project, (415) 346-8223, www.breastcancerfund.org.

Why do modern bacteria "resist" antibiotics, confounding medical treatment?

Hugo Mestres, Seattle, WA

Antibiotics have played a profoundly important role in staving off bacterial infections since Alexander Fleming first discovered them in 1927. But the effectiveness of these so-called miracle drugs has waned in recent years, because some of the very bacteria they are meant to control are mutating into new forms that don't respond to treatment. Many medical experts blame this phenomenon on both the misuse and overuse of antibiotics in recent years, in both human medicine and in agriculture.

Doctors first noticed antibiotic resistance more than a decade ago when children with middle ear infections stopped responding to them. Penicillin as a treatment for strep has also become increasingly less effective. And a recently discovered strain of staph bacteria does not respond to antibiotic treatments at all, leading medical analysts to worry that certain "superbugs" could emerge that are resistant to even the most potent drugs, rendering some infections incurable. The Centers for Disease Control and Prevention (CDC) calls antibiotic resistance one of its "top concerns" and "one of the world's most pressing health problems."

One large part of the problem, according to the CDC, is the tendency for people to take antibiotics to fight viruses. Antibiotics fight bacteria, not viruses, and won't cure colds, flu, bronchitis, runny noses, or sore throats not related to strep. Despite that, the Centers for Disease Control says that "more than 10 million courses of antibiotics are prescribed each year for viral conditions that do not benefit from antibiotics." To address this, a growing number of doctors, including Dr. Randel Cardott, an internist with Iowa's Genesis Convenient Care, are advocating a wait-and-see approach to prescribing antibiotics, especially in cases like middle ear infections that sometimes prove to be viral and not bacterial in origin. Cardott says that European physicians have taken this approach for years with no adverse effects.

Scaling back on human antibiotic use won't solve the problem. Farmers and ranchers also use antibiotics heavily. In North America, industrial beef, pig, and poultry farming is a big, unsanitary business, and antibiotics are used extensively to ward off diseases and also for nonmedical reasons, such as to promote growth. In fact, the Union of Concerned Scientists (UCS) estimates that some 70 percent of all antibiotics are used as additives in the feed given to healthy pigs, poultry, and cattle. These drugs leave the animals' bodies as waste and work their way into local water supplies as well as right into the food chain. "Nonetheless," says UCS, "agribusiness and the pharmaceutical industry are fighting hard to thwart restrictions on the use of antibiotics in agriculture."

The group Keep Antibiotics Working advocates phasing out unnecessary antibiotics in healthy livestock and poultry. In lieu of congressional action along these lines, the group is encouraging meat wholesalers and retailers to voluntarily stop buying or selling meat that has been produced using antibiotics for purposes other than treating sick animals. Consumers looking to avoid antibiotics in meat should seek out organic offerings at natural foods markets.

CONTACTS: Keep Antibiotics Working, www.keepantibioticsworking.com; Union of Concerned Scientists, www.ucsusa.org/food_and_environment/antibiotics_and_food.

I recently heard the term "conservation medicine." What is it exactly?

Steve Falbo, San Francisco, CA

Conservation medicine (sometimes called "conservation health") is a relatively new field that studies the links between human health, animal health, and the environment. It specializes in the recent emergence of deadly diseases that have crossed over from animals to humans, including mad cow, AIDS, Lyme disease, SARS, avian flu, and West Nile virus. Many of these plagues arose out of some form of human-animal contact in compromised ecosystems.

In 1998, for example, a previously unknown virus spread among some farm families in Malaysia, eventually killing more than one hundred people. The outbreak was traced back to a pig farm where horses, cats, dogs, and goats were also infected. The virus, named Nipah for one of its first human victims, eventually spread to Singapore, where nine slaughterhouse workers became ill after processing Malaysian pig meat.

Scientists concluded that the virus came from fruit bats that descended on Malaysia after their native habitat, forests in nearby Borneo and Sumatra, had been clear-cut. The bats sought refuge in the fruit trees hanging over the animal pens at the pig farm, and then passed the virus to the pigs by dropping infected fruit into the pens—to be eagerly consumed by the pigs. How the virus jumped to humans is still a mystery, but scientists are quite sure that the clearing of forests in Borneo and Sumatra indirectly led to more than one hundred human deaths.

"Diseases are moving from animals to humans and from one animal species to another at an alarming rate," says Lee Cera, a veterinarian at Loyola University's Stritch School of Medicine. "When I went to school we were told, 'This disease won't go from a dog to a cat.' Then all of a sudden a dog virus wiped out the lions of the Serengeti. How did it happen? When did it happen?" Conservation medicine is an attempt to answer these questions by bringing together professional fields that had previously worked in isolation: human medicine, veterinary medicine, infectious disease research, public health, and environmental science.

Increased human forays into wilderness areas (often spurred by population growth) have set up new points of human-animal contact. The international trade in exotic species also breaks down previously existing barriers. Climate change causes species to migrate to new areas, bringing with them new germs. Global travel plays a role: In 1950, three million people flew on commercial jets. In 1990, three hundred million did. Two million people now cross international borders daily, carrying with them huge amounts of agricultural products, live animals, soil, and disease-causing microbes.

At the forefront of the new field are two groups based in New York: the Wildlife Trust and the Consortium for Conservation Medicine. "Conservation medicine demonstrates how healthy ecosystems are the basis for human well-being," says Mary Pearl, Wildlife Trust's president. "And it can really engage people who didn't see the relevance before."

CONTACTS: Consortium for Conservation Medicine, www.conservationmedicine.org; the online journal *Environmental Health Perspectives*, www.ehponline.org; Wildlife Trust, www.wildlifetrust.org.

What happens to the chemicals in drugs once they are out of our systems?

Courtney Moschetta, Huntsville, AL

Every time you swallow a pill, some of that medicine follows a circuitous path through your body, down the toilet, through the sewage-treatment plant (whose efforts it can and will resist), and into the nearest river or lake, where it is eventually tapped again for the public drinking water supply.

According to Christian Daughton, chief of environmental chemistry at the National Exposure Research Laboratory in Las Vegas, new technologies now allow scientists to detect extremely low levels of prescription and over-the-counter drugs, as well as compounds found in personal-care products like shampoo and sunscreen, in water. In Kansas City alone, more than 40 percent of stream samples analyzed recently by the U.S. Geological Survey had detectable amounts of over-the-counter drugs like ibuprofen and acetaminophen, antibiotics, and prescription medications for high blood pressure.

While the effects on human health are not yet getting much attention, new studies show that fish and other aquatic species may be affected, says Daughton. Antibiotics make some species more resistant to pathogens, steroids can cause endocrine disruption that interferes with reproductive processes, and antidepressants make fish tranquil and more likely to succumb to predation. Considering the large variety of pharmaceuticals on the market today, our water may have a witches' brew of very small amounts of many different kinds of drugs.

Right now there are no EPA or FDA regulations in place to control levels of residual drugs in water, but some environmental groups want to see drug-disposal policies enacted, new sewage-treatment technologies developed, and source reduction efforts on the part of pharmaceutical companies and pharmacies. Daughton envisions a day when drug companies will take responsibility for the life cycle of their products. Instead of flushing your unused prescription drugs down the toilet, you may be able to send them back to the pharmacy or return them to the maker for proper disposal. Not surprisingly, such programs already exist in Europe and Canada.

CONTACT: EPA National Exposure Research Laboratory, Environmental Sciences Division, www.epa.gov/ppcp; U.S. Geological Survey's Toxic Substances Hydrology Program, http://toxics.usgs.gov.

Are there any safe herbal alternatives to common pain and headache medications?

Dylan Baker, Philadelphia, PA

From lower back problems and migraines to the stings of cuts, bruises, and more serious injuries, persistent pain sends most of us running to the doctor—and then the pharmacist—for a quick fix. Often that relief comes in the form of a potent pain pill, such as OxyContin or Percocet. But those drugs exact a price, over and above their financial cost, to work their magic. For one, they can be highly addictive. And many have serious side effects ranging from drowsiness to ulcers to kidney and liver damage.

New research shows that many of these powerful pharmaceuticals are also no friends to the environment. When they are eliminated from our bodies and flushed down the toilet, they make their way into our waterways and dissolve into microscopic particles. Fish and wildlife living in and near streams polluted by these compounds can develop health problems. So can people who drink the water from public water supplies whose treatment methods are not sophisticated enough to screen out the particles.

Given those risks, millions of American families have turned to herbal pain remedies. Pain relief patches are among the hot new products on display at natural foods markets. Some, such as the Tiger Balm patch, contain the herbs camphor and oil of clove, which are absorbed through the skin. These herbs, originally used in Chinese medicine, have anti-inflammatory properties to ease back pain and muscle aches.

Arnica, also known as leopard's-bane, is especially effective in reducing pain from arthritis, burns, ulcers, and eczema and is also used to treat acne. It has antibacterial and anti-inflammatory qualities that reduce pain and swelling and accelerate wound healing. According to the American Botanical Council, arnica is one of the most popular herbs used by homeopathic practitioners for pain.

Meanwhile, several research studies have shown that capsaicin, an extract of cayenne pepper, can offer significant arthritis pain relief when used as a topical cream. Turmeric, a natural anti-inflammatory and immune booster, is also well known as a pain reliever for arthritis pain. And feverfew is a useful herb for reducing migraine headache pain. Herbalists have relied on it since the Middle Ages.

Since herbal remedies are not as strictly regulated as conventional drugs, it is important to consult a reputable naturopath or homeopath before using them. For example, despite its benefits, arnica can raise blood pressure and therefore may be undesirable for some people. Also, research shows a wide variation in purity and quality among the herbal offerings on the market, so it is also important to choose a reputable manufacturer. Websites such as mothernature.com and medicine-plants.com are good sources as are the herb and supplement sections in Wild Oats, Whole Foods, and other natural foods markets.

Regardless of where you obtain herbal medicines, be aware that many herbs can interact with other medicines, so you should check with your medical doctor before using them. Also, as with conventional medicines, it may also be unwise to take herbal treatments if you are pregnant or trying to conceive.

CONTACTS: American Botanical Council, www.herbalgram.org; Medicine Plants, www.medicine-plants.com; Mother Nature, Inc., www.mothernature.com.

Which types of household products are most likely to cause chemical sensitivities?

John Morgan, Somerville, MA

The answer is probably under your sink. Household products trigger chemical sensitivities in hundreds of thousands of Americans every year, yet few people make the connection between their skin rash or sneezing and common cleaning products.

Common reactions to everyday household cleansers and other substances include migraine headaches, asthma, and sinusitis, but more serious cardiovascular, neurological, and autoimmune diseases may also result from prolonged use or lack of adequate ventilation in areas where these chemicals are being applied. "Early warning signs are burning and irritation of the sinuses, nose, or throat—usually not with a fever—and itching or sneezing," says Dr. Grace Ziem, a public health physician specializing in chemical injuries.

Prevention is the key. And removing toxic compounds from your home is the strategy. You can begin under the kitchen sink by replacing traditional choices with "products your grandmother bought," says Suzanne Olson of the Environmental Health Network. "Borax, vinegar, and baking soda will clean most items around the house." Olson uses vegetable oil to polish furniture and shuns any items with a fragrance.

"If any ingredients end in 'ethylene' or 'ethane,' it's not a healthy product," says Cynthia Wilson of the Chemical Injury Information Network. She recommends using scent-free and dye-free laundry products and oxygen-based whitening additives in place of toxic bleach. Two companies that supply nontoxic laundry products as well as other green-friendly household cleaners include Seventh Generation and Earth Friendly Products. You can shop for both online or in natural foods markets and some supermarkets.

Synthetic home furnishings can also trigger sensitivities. Foam, particleboard, and veneers can all aggravate a variety of symptoms. And sweet dreams may elude you if you have chemical sensitivities to items in your bedroom. Most mattresses are made from artificial materials, and some beds have chemical mold inhibitors while almost all have been treated with a fire retardant. In order to eliminate chemical sensitivities from ruining your sleep, choose an organic cotton mattress. Two good sources include Lifekind Products and Heart of Vermont; both offer secure online ordering.

If you think household chemicals might be bothering you, Dr. Ziem suggests keeping a log to help pinpoint the offenders. The best way to find out whether any chronic ailments you may have are caused or aggravated by household products is to take a simple inventory of the home and its contents and replace synthetic products with natural ones wherever possible.

CONTACTS: Chemical Injury Information Network, www.ciin.org; Earth Friendly Products, (800) 335-3267, www.ecos.com; Environmental Health Network, http://users.lmi.net/wilworks; Heart of Vermont, (800) 639-4123, www.heartofvermont.com; Lifekind, www.lifekind.com, (800) 284-4983; Seventh Generation, (800) 456-1191, www.seventhgeneration.com.

Are there organic highlights and dyes I can use in my hair that contain less ammonia and peroxide than traditional brands?

Terry Wattendorf, Scituate, MA

We all have one life and might want to live it as a blonde, but the harsh, allergy-aggravating chemicals in popular dyes and highlighting treatments are enough to give anyone pause. Green-friendly permanent hair dye alternatives do have some of these chemicals—including ammonia, peroxide, p-phenylenediamine, or diaminobenzene—to be effective, but the best of them contain very small amounts.

EcoColors, with small amounts of ammonia and peroxide, has a soy and flax base and uses rosemary extract to condition the hair and flower essences instead of artificial scents. Another option is Herbatint. This ammonia-free permanent dye is biodegradable, but it does use low concentrations of p-phenylenediamine and peroxide.

Temporary dyes and highlight treatments color hair without the need for harsh chemicals. Naturcolor and Vegetel are shorter lived options that do not contain any damaging chemicals, although their effect will only last for a few washes.

One truly natural although temporary dye that has been around since Cleopatra herself is henna. Made from the powdered leaves of a desert shrub called *Lawsonia inermis*, henna has been used for thousands of years to color hair and skin. Rainbow Henna makes a variety of 100 percent organic hair treatments ranging from blonde to black and including everything in between. Meanwhile, Light Mountain sells an organic henna application kit familiar to those accustomed to traditional home hair-coloring packages. While many such treatments are available at natural health and beauty supply retailers, others, such as the Italian-made Tocco Magico, may be available only at salons.

Recent studies add to the worries about traditional hair dyes. One report found that deep-colored dyes (like dark brown and black), when used over prolonged periods of time, seem to increase the risk of cancers such as non-Hodgkin's lymphoma and multiple myeloma. Another study in 2001 found that people who use permanent hair dyes are twice as likely to develop bladder cancer as those who go au naturel regarding hair color.

CONTACT: FDA, Office of Cosmetics and Colors, www.cfsan.fda .gov/~dms/cos-toc.html.

Are there any environmental or human health risks to using nail polish?

Deborah Lynn, Milford, CT

The evidence is fairly convincing. Conventional nail polishes dispensed at most drugstores and nail salons contain a veritable witches' brew of chemicals, including toluene, which has been linked to a wide range of health issues from simple headaches and eye, ear, nose, and throat irritation to nervous system disorders and damage to the liver and kidneys.

Another common yet toxic ingredient in conventional nail polish is a chemical plasticizer known as dibutyl phthalate (DBP). According to the Environmental Working Group (EWG), a nonprofit research and advocacy organization that campaigns to educate consumers about the health risks of cosmetics, studies have linked DBP to underdeveloped genitals and other reproductive system problems in newborn boys.

DBP is banned from cosmetics in the European Union, but the FDA has taken no such action in the United States, though a recent study found DBP and other toxic phthalates in the bloodstreams of every person they tested. Further, 5 percent of women tested who were of childbearing age (between twenty and forty) had up to forty-five times more of the chemicals in their bodies than researchers had expected to find.

EWG attributes the prevalence of DBP in young women to widespread use of nail polish. "Women of childbearing age should avoid all exposure to DBP when they're considering becoming pregnant, when they're pregnant, or when they're nursing," says Jane Houlihan, an EWG vice president.

Luckily, safer nail polishes do exist and are readily available at natural health and beauty supply stores as well as from online outlets such as Natural Solutions and Infinite Health Resources. These products—from such makers as Honeybee Gardens, PeaceKeeper, Jerrie, Visage Naturel, and Sante—rely on naturally occurring minerals and plant extracts to beautify nails without the need for toxic ingredients.

Major nail polish manufacturers are also now getting in on the act. According to the Campaign for Safe Cosmetics, a coalition of organizations that includes EWG and the Breast Cancer Fund, Avon, Estée Lauder, Revlon, and L'Oréal confirmed last year that they would begin removing DBP from products. And leading drugstore brand Sally Hansen has said it is reformulating all of its products to remove DBP and toluene as well as formaldehyde, which is also known to cause cancer and reproductive problems.

Exposure to toxic chemicals is not the only health concern associated with nail salons, where nail fungus and bacteria can lurk on the underside of any emery board. Women's health advocate Tracee

Cornforth suggests checking out a salon for cleanliness before signing up for services. She also says to make sure attendants disinfect all tools and equipment between customers and even recommends bringing in one's own manicure or pedicure kit so as to minimize the transmission of any unsightly or painful maladies.

CONTACTS: Campaign for Safe Cosmetics, www.safecosmetics.org; Environmental Working Group, www.ewg.org; Infinite Health Resources, www.infinitehealthresources.com; Natural Solutions, www.bewellstaywell.com.

A number of "natural" shampoos claim to be "sodium lauryl sulfate free." What is sodium lauryl sulfate, and should it be avoided?

Kristen Lohse, Seattle, WA

Sodium lauryl sulfate (SLS) is a synthetic detergent known for its ability to generate a sudsy lather. As a result, the beauty and cosmetics industry has long used it as a key component in shampoos and other personal-care products, because foam matters.

But what America's happy bathers don't know is that SLS dries the scalp, stripping the skin's surface of its protective lipids. Follicle damage, hair loss, skin and eye irritation, and allergic reactions such as rashes and hives can result. And if accidentally ingested, SLS can lead to gastrointestinal and/or liver distress.

Despite these potential maladies, nine out of ten shampoo brands contain SLS or one of its variants. And since the FDA does not regulate beauty or cosmetics products, SLS is likely to remain right where it is as long as consumers want sudsy shampoos.

If you switch to SLS-free shampoo, you might miss the sudsy lather you've grown accustomed to in mass-marketed products, but you can still expect clean and manageable hair. Manufacturers of all-natural shampoos usually opt for good old-fashioned soap instead of SLS-based detergent to get the cleaning done. "It is a fallacy that you need to have foaming bubbles to get it clean," says Dr. Ron Shelton of the American Academy of Dermatology.

Buyer beware, though: not all shampoos marketed as "natural" or "organic" are SLS free. Check ingredients lists for SLS or variations such as sodium laureth sulfate (SLES), ammonium lauryl sulfate (ALS), or ammonium laureth sulfate (ALES). All-natural herbal shampoos are least likely to contain SLS-type products. Aubrey Organics, Aveda, and Kiss My Face, among many other companies, make SLS-free shampoos. Check in your local natural foods markets.

CONTACTS: American Academy of Dermatology, (866) 503-7546, www.aad.org; Aubrey Organics, (800) 282-7394, www.aubrey-organics.com; Aveda, (866) 823-1425, www.aveda.com; Kiss My Face, (800) 262-KISS (5477), www.kissmyface.com.

Are contact lens fluids safe for the environment and personal health?

M. Luh, Storrs, CT

Now there's a question we need to, well, look at. According to the Contact Lens Council, approximately thirty-four million Americans now wear contact lenses. Most people use saline and disinfectant solutions from big-name companies like Bausch & Lomb and Johnson & Johnson. These products are usually packaged in squeeze bottles that, according to FDA regulations, are required to have preservatives.

But many preservatives can cause irritation and discomfort to the user. For example, thimerosal, a preservative commonly used in contact lens disinfectant solutions in the past, was found to be the culprit in severe allergies. One particularly harmful by-product of thimerosal, once it degrades or metabolizes, is ethyl mercury, which researchers now believe plays a role in the development of autism and a number of health problems.

Other preservatives like benzalkonium chloride have now replaced thimerosal in contact lens solutions. However, according to Switzerland-based Opticiens Perret, there is still the possibility of such allergic reactions as redness, itching, and discharge with the new generation of preservatives being used in conventional solutions.

There are other options. Clear Conscience now makes a line of contact lens solutions that is reportedly gentler for the environment and the wearer. Unlike conventional products, the company's saline solution is dispensed by a safe nitrogen propellant, which means it contains no preservatives. Clear Conscience also offers an FDA-approved multipurpose solution that is benzalkonium chloride and thimerosal free. It can be used to disinfect, clean, and store hard or soft contacts. Whole Foods Market and other large natural goods stores stock Clear Conscience products, or they can be ordered online at the company's website.

Alcon, one of the largest providers of traditional lens solutions, has diversified its product line to include two preservative-free products, Unisol 4 and Pliagel, both of which are available online at drugstore .com, as well as at other large health-products retailers.

CONTACTS: Alcon, (800) 862-5266, www.alconlabs.com; Clear Conscience, (800) 595-9592, www.clearconscience.com; Contact Lens Council, (800) 884-4CLC, www.mycontactlenses.org; drugstore.com, inc., www .drugstore.com; Opticiens Perret, www.perret-optic.ch/index_gb.htm.

How serious is the threat of antibiotic-resistant bacteria in chicken and other poultry?

Dana Wilke, Chicago, IL

You should indeed take the threat seriously. According to the Union of Concerned Scientists, 70 percent of all antibiotics in the United

States are fed to pigs, cattle, and poultry so they'll grow quickly and stay healthy. Meanwhile, humans rely on many of these same antibiotics as medicines to control various bacterial infections. Bacteria in poultry and other livestock exposed over and over to these antibiotics develop increased resistance. The result can be that when people become infected by these same bacteria—such as campylobacter or salmonella, the two most common causes of food poisoning in the United States—the antibiotics they normally rely on can be useless.

The Keep Antibiotics Working (KAW) campaign—an association of health, consumer protection, environmental, and animal welfare organizations—says that antibiotic resistance is "reaching crisis proportions, resulting in infections that are difficult, or impossible, to treat." According to KAW, "Overuse and misuse of antibiotics greatly accelerates the proliferation of resistant bacteria."

A recent study cited in *Consumer Reports* found that 49 percent of brand-name whole broiler chickens purchased in food stores in twenty-five U.S. cities were contaminated with campylobacter and/or salmonella bacteria. According to KAW, those two strains of bacteria alone cause 3.3 million illnesses and 650 deaths every year. The study also found that 90 percent of the campylobacter and 34 percent of the salmonella tested were resistant to at least one antibiotic.

The Institute for Agriculture and Trade Policy (IATP) found that thousands of people in the Minneapolis area were ingesting bacteria resistant to important antibiotic medicines like Cipro, Synercid, and tetracycline. "As bacteria on food get more and more resistant to the antibiotics doctors rely on for treating infections, it puts patients' lives at risk. This study confirms that supermarket chicken . . . can be an important source of drug-resistant infections," says IATP's Dr. David Wallinga. "We can't afford to play Russian roulette with our existing antibiotics because they are rapidly losing effectiveness," he concludes.

CONTACTS: Institute for Agriculture and Trade Policy, (612) 870-0453, www.iatp.org; Keep Antibiotics Working, (773) 525-4952, www

.keepantibioticsworking.com; Union of Concerned Scientists, (617) 547-5552, www.ucsusa.org.

I saw warnings on bags of charcoal that said carcinogens are released when the briquettes are burned. Is it safe to breathe in the smell of a charcoal grill?

Joe Sliwa, via e-mail

Here's a paradox: the very thing we love about charcoal grilling—that delicious aroma—is what's dangerous to us. Barbecue grills can be problematic because both charcoal and wood burn "dirty," producing not only hydrocarbons but also tiny soot particles that pollute the air and can aggravate heart and lung problems. Also, the grilling of meat can form two kinds of potentially carcinogenic compounds: polycyclic aromatic hydrocarbons (PAHs) and heterocyclic amines (HCAs).

PAHs form when fat from meat drips onto the charcoal. They then rise with the smoke and can get deposited on the food. They can also form directly on the food as it is charred. The hotter the temperature and the longer the meat cooks, the more HCAs are formed.

HCAs can also form on broiled and pan-fried beef, pork, chicken, and fish, not just on grilled red meat. Researchers have identified seventeen different HCAs that result from cooking "muscle meats" and that may pose human cancer risks. Studies have also shown increased risk of colorectal, pancreatic, and breast cancers associated with high intakes of well-done, fried, or barbecued meats.

Texans like to say that they "live and breathe barbecue," but they may be doing just that to the detriment of their health. A 2003 study by scientists from Rice University found that microscopic bits of polyunsaturated fatty acids released into the atmosphere from cooking meat on backyard barbecues are helping to pollute the air in Houston. The city recently surpassed Los Angeles with the worst air quality in the United States, though emissions from barbecues take a big back-seat to car exhaust as the culprit.

Both briquettes and lump charcoal create air pollution. Lump charcoal, made from charred wood to add flavor, also contributes to deforestation and adds to the greenhouse gases in the atmosphere. Charcoal briquettes do have the benefit of being made partly from sawdust (a good use of waste wood), but popular brands may also contain coal dust, starch, sodium nitrate, limestone, and borax.

In Canada, charcoal is now a restricted product under the Hazardous Products Act. According to the Canadian Department of Justice, charcoal briquettes in bags that are advertised, imported, or sold in Canada must display a label warning of the potential hazards of the product. No such requirements presently exist in the United States.

You can avoid exposure to these potentially harmful additives by sticking with so-called natural charcoal brands. Noram de México's Sierra Madre 100 percent oak hardwood charcoal contains no coal, oil, limestone, starch, sawdust, or petroleum products and is also certified by the Rainforest Alliance's SmartWood program as sustainably harvested. The product is available at select Sam's Clubs across the United States. Other manufacturers of all-natural charcoal include Greenlink and Lazzari, both of which can be found at natural food outlets.

CONTACTS: Greenlink Charcoal, www.greenlinkcharcoal.com; Lazzari, www.lazzari.com; Rainforest Alliance SmartWood program, www.rainforest-alliance.org/programs/forestry/smartwood.

Are "silver" dental fillings, which contain mercury, toxic?

Erin Stills, Miami, FL

Despite the name, "silver" fillings are actually composed of about 50 percent mercury and 30 percent silver, with the rest made of copper, tin, zinc, and sometimes cadmium. A vocal minority of Americans complain that the fillings have damaged their health through mercury poisoning, causing shortness of breath, loss of energy, memory damage, and even partial paralysis.

Also called amalgam, silver fillings are cheap and easy to install, and the American Dental Association (ADA) reports that 76 percent of dentists use them. Although the ADA concedes that "a very small number of people" are allergic to the fillings, the group staunchly maintains, "Studies have failed to find any link between amalgam restorations and any medical disorder."

The ADA has long claimed that mercury remains chemically locked within the "extremely stable" fillings, but according to the U.S. Agency for Toxic Substances and Disease Registry, "Very small amounts are slowly released from the surface of the filling due to corrosion or chewing or grinding motions." Although the agency agrees with the ADA that there is not yet scientific agreement on whether this exposure actually does cause health problems, it suggests that fillings may be risky for pregnant women, children, and those with impaired kidney or immune function.

The citizen group Consumers for Dental Choice argues that mercury fillings do pose a significant threat to public health, and its members are campaigning to end the practice.

CONTACT: American Dental Association, www.ada.org; Consumers for Dental Choice, (202) 544-6333, www.toxicteeth.org.

Are there toothpastes on the market that don't contain chemicals or artificial sweeteners?

Jeffrey Moss, Westport, CT

You bet, but you'll have to seek them out. Most conventional toothpastes use saccharin as a sweetener. Although it has not been proven that saccharin causes cancer in humans, many studies have linked it to cancer in laboratory animals, and some experts, including Dr. Samuel Epstein of the University of Illinois Medical Center, recommend that consumers avoid it.

Fluoride has also come under fire in recent years because of its suspected ties to bone cancer, hip fractures, and fluorosis (white spots and blotching on teeth caused by excessive ingestion of fluoride). Although the American Dental Association (ADA) strongly endorses fluoride-containing products, claiming they are safe and effective for cavity prevention, some experts argue that if fluoride can damage tooth-forming cells, as in fluorisis, then other harm to the body may also occur.

Triclosan is the most often used antibacterial agent in toothpaste. The EPA considers triclosan a pesticide and a chlorophenol, part of a class of chemicals thought to cause cancer in humans. Sodium lauryl/laureth sulfate, a foaming agent, and sorbitol are two other oral hygiene ingredients whose safety has been questioned. And most so-called whitening toothpastes use sodium or potassium hydroxides, also known as lye, considered a poison by the FDA.

For many years the alternatives to mass-market toothpastes were plain baking soda, tooth powders, or bad-tasting pastes that most adults disliked and kids refused to use. There are many new pastes on the market now that, if somewhat less sweet-tasting than those with saccharin, taste great—and the dental establishment is warming up to them.

The ADA has awarded its seal to Tom's of Maine, which makes a large variety of natural-ingredient toothpastes. And the *Journal of Clinical Dentistry* found that Herbal Toothpaste and Gum Therapy from The Natural Dentist outperformed Colgate's Total in reducing gingivitis and teeth stains. The Natural Dentist makes pastes and gels in a variety of flavors that contain sodium laureth sulfate, but don't use artificial sweeteners, preservatives, or dyes. Peelu Toothpaste, which comes in spearmint, cinnamon, and peppermint flavors, uses *peelu*, a vegetable fiber, as an abrasive and glycerine as a cleanser, rather than a synthetic detergent. Weleda makes toothpaste free of saccharin and sodium lauryl sulfate. Its Pink Toothpaste with myrrh contains nine essential oils for gum health, and its Children's Tooth Gel is made especially for young teeth.

For consumers wanting to avoid fluoride, Tom's of Maine makes fluoride-free natural toothpaste for adults and children. Tom's also makes a whitening toothpaste that uses silica; Jason Natural makes one that uses both silica and bamboo powder.

CONTACTS: American Dental Association, (312) 440-2500, www.ada.org; Jason Natural, (877) JASON-01, www.jason-natural.com; Peelu Toothpaste, (701) 356-2811, www.peelu.com/peelutoothpaste.html;

The Natural Dentist, (800) 615-6895, www.thenaturaldentist.com; Tom's of Maine, (800) 367-8667, www.tomsofmaine.com; Weleda, (800) 241-1030, www.usa.weleda.com.

How serious is the risk of contracting Alzheimer's disease from using antiperspirants that contain aluminum?

Susan DeBacker, Boulder, CO

We go to great lengths to avoid offending others, perhaps dangerous lengths. Antiperspirants often contain aluminum, zirconium, or both. These substances tighten or close underarm skin pores in order to block sweat glands and the moisture they produce. While underarm products often contain both antiperspirants and deodorants, deodorants alone do not contain aluminum.

Could exposure to aluminum increase your chances of getting Alzheimer's disease? According to the Alzheimer's Disease Education and Referral Center, Alzheimer's patients do at times have abnormally high concentrations of aluminum in their brains, but research hasn't conclusively shown if the disease causes the buildup or the buildup causes the disease. Some doctors have suggested that antiperspirant might be especially problematic, as women apply it to shaved armpits, perhaps allowing aluminum to be absorbed directly into the bloodstream through hundreds of tiny cuts caused by razors. However, studies are inconclusive. "The research hasn't shown anything further regarding the link between Alzheimer's and aluminum," says Jennifer Watson, outreach and promotions specialist at the center.

Procter & Gamble, which makes antiperspirants with aluminum, points out that aluminum is Earth's third most common element, and that humans are routinely exposed to it through numerous sources besides antiperspirants, including tainted water, canned foods, processed cheese, and buffered aspirin.

A simple precaution is to buy deodorants without antiperspirants. Some companies, such as Nature's Gate, offer deodorants that fight odor-causing bacteria with all-natural ingredients.

CONTACT: Alzheimer's Disease Education and Referral Center, (800) 438-4380, www.alzheimers.org; Nature's Gate, (800) 327-2012, www.natures-gate.com.

With the recent hubbub over the chemicals used to make Teflon linked to health problems, what is the safest cookware to use in preparing meals for my family?

Wyatt Walley, Needham, MA

When the health risks associated with making Teflon first came to light, many cooks trashed their nonstick cookware and went back to using their old stainless steel pots and pans. But what many people didn't realize was that even stainless steel is not immune to health controversies.

In fact, stainless steel is really a mixture of several different metals, including nickel, chromium, and molybdenum, all of which can trickle into foods. However, unless your stainless steel cookware is dinged and pitted, the amount of metals likely to get into your food is negligible.

These days, many health-conscious cooks are turning to anodized aluminum cookware as a safer alternative. The electrochemical anodizing process locks in the cookware's base metal, aluminum, so that it can't get into food and makes for what many cooks consider an ideal nonstick and scratch-resistant cooking surface. Calphalon is the leading manufacturer of anodized aluminum cookware, but newer offerings from All-Clad (endorsed by celebrity chef Emeril Lagasse) and others are coming on strong.

Another good choice is that old standby, cast iron, which is known for its durability and even heat distribution. Cast-iron cookware can also help ensure that eaters in your house get enough iron—which the body needs to produce red blood cells—as it seeps off the cookware into food in small amounts. Unlike the metals that can come off some other types of pots and pans, iron is considered a healthy food additive by the FDA. But most cast-iron cookware needs to be seasoned after each use and as such is not as worry free as other alternatives. Lodge Manufacturing is the leading American producer of cast-iron cookware.

If you like the feel and heat distribution properties of cast iron but dread the seasoning process, ceramic enameled cookware from Le Creuset, World Cuisine, and others is a good choice. The smooth and colorful enamel is dishwasher friendly, somewhat nonstick, and covers the entire surface to minimize cleanup headaches. One other surface favored by chefs for sauces and sautés is copper, which excels at quick warm-ups and even heat distribution. Since copper can leak into food in large amounts when heated, the cooking surfaces are usually lined with tin or stainless steel.

But don't trash your nonstick cookware just yet. According to DuPont, the finished product of Teflon does not contain any of the production-process chemicals linked to health problems in factory workers. And the EPA says that ingesting small particles of Teflon flaked off into food is not known to cause any health maladies. With proper use and care, such pots and pans—which constitute more than half of all cookware sales in the United States—should be safe to use for years to come.

CONTACTS: Calphalon, www.calphalon.com; All-Clad, www.allclad.com; Le Creuset, www.lecreuset.com; Lodge Manufacturing, www.lodgemfg.com; World Cuisine, www.world-cuisine.com.

Why is chlorine added to tap water? Do water filters effectively filter it out?

J. P. Miller, Hudson, WI

It's not there as part of some vast right-wing conspiracy. Chlorine is a highly efficient disinfectant and is added to public water supplies to kill disease-causing bacteria that the water or its transport pipes might contain. "Chlorine has been hailed as the savior against cholera and various other waterborne diseases, and rightfully so," says Steve Harrison, president of water-filter maker Environmental Systems Distributing. "Its disinfectant qualities . . . have allowed communities and whole cities to grow and prosper by providing disease-free tap water to homes and industry."

But Harrison says that all this disinfecting has not come without a price: chlorine introduced into the water supply reacts with other naturally occurring elements to form toxins called trihalomethanes (THMs), which eventually make their way into our bodies. THMs have been linked to a wide range of human maladies ranging from asthma and eczema to bladder cancer and heart disease. Dr. Peter Montague of the Environmental Research Foundation cites several studies linking moderate to heavy consumption of chlorinated tap water by pregnant women with higher miscarriage and birth defect rates.

The Environmental Working Group concluded that from 1996 to 2001, more than sixteen million Americans consumed dangerous amounts of contaminated tap water. The report found that water supplies in and around Washington, D.C., Philadelphia and Pittsburgh in Pennsylvania, and the Bay Area in California were putting the greatest number of people at risk.

"Dirty water going into the treatment plant means water contaminated with chlorination by-products coming out of your tap," said Jane Houlihan, EWG's research director. "The solution is to clean up our lakes, rivers, and streams, not just bombard our water supplies with chlorine."

Eliminating water pollution and cleaning up our watersheds is not going to happen overnight, but alternatives to chlorination for water treatment do exist. Dr. Montague reports that several European and Canadian cities now disinfect their water supplies with ozone instead of chlorine. Currently a handful of U.S. cities do the same, most notably Las Vegas, Nevada, and Santa Clara, California.

Those of us who live far from Las Vegas or Santa Clara do have other options. First and foremost is filtration at the faucet. Carbon-based filters are considered the most effective at removing THMs and other toxins. The consumer information website waterfilterrankings.com compares various water filters on the bases of price and effectiveness. The site reports that filters from Paragon, Aquasana, Kenmore, GE, and Seagull remove most if not all of the chlorine, THMs, and other potential contaminates in tap water.

If you don't have money to spend on home filtration, you can just rely on good old-fashioned patience. Chlorine and related compounds will make their way out of tap water if the container is simply left uncovered in the refrigerator for twenty-four hours.

CONTACTS: Environmental Research Foundation, www.rachel.org; Environmental Working Group, www.ewg.org; Waterfilterrankings.com, www.waterfilterrankings.com.

4
LIVING (AND WORKING) SPACES

Digging In, Turning Off, and Waxing On

We all know we need to get outside more. We spend mornings in our cars navigating traffic, remain seated at an office for most of the day, eating sandwiches in front of computer screens, and then inch our way back home in our cars to finally flop, exhausted, in front of our television sets. Even our New Year's resolutions tend only to lead us (temporarily) to another building, the warehouselike gym, to burn off steam. While it would be wonderful to hop off the hamster wheel and begin revisiting the great outdoors, there are a lot of things we can do in the meantime to bring green into the indoor environments where we spend the better part of our days. We can grow more plants and use less water, minimize our paper and disposable cup waste at the office, and give some thought to the myriad chemical cleaners we use to spray and wipe the surfaces that make up our homes and cubicles. Once we start breathing cleaner air inside, we might just be inspired to don our walking shoes.

What is sick building syndrome?

Annie Sundberg, New York, NY

Just like human illnesses, sick buildings need time and treatment to heal. The term "sick building syndrome" (SBS) was coined in the 1970s when researchers first linked a rash of office outbreaks to the environmental conditions in their work spaces.

According to the EPA, the afflicted can experience headaches; eye, nose, or throat irritation; dry cough; dry or itchy skin; dizziness and nausea; difficulty concentrating; fatigue; and extra sensitivity to odors.

Usually, sick building syndrome is associated with commercial buildings, but residential homes can also trigger symptoms. And, according to the U.S. Green Building Council, more than half of all U.S. schools have sick building syndrome.

Ironically, improvements in building design and energy efficiency may be major contributors to the problem, as airtight indoor space is not as well ventilated as drafty space. According to the National Institute for Occupational Safety and Health, indoor air pollution, biological contaminants such as bacteria and mold, and inadequate ventilation have all contributed to a rise in SBS in recent years.

Adhesives, upholstery, carpeting, copiers, manufactured wood products, cleaning agents, and pesticides are all sources of indoor air pollution, as are many of the chemical smells and other odors present in manufacturing and service settings. Also, according to the EPA, outdoor pollutants such as car exhaust can enter buildings through poorly located air-intake vents and windows and become trapped indoors.

It's no wonder that sick building syndrome has been on the rise in recent years: People are spending more and more time indoors, and building materials, furniture, and equipment contain many more synthetic chemicals than they did fifty years ago. Buildings operated or maintained in ways they were not originally designed for can create problems, as can our own habits, such as smoking or the use of colognes and perfumes.

Not everyone is convinced that on-the-job illnesses are associated with a building's environmental factors. In a study conducted by Dr. Mai Stafford of the University College London Medical School, symptoms were strongly linked to other factors, such as job stress and a lack of social support at work. Dr. Stafford and colleagues concluded that "if sick building syndrome is reported in a building, management should consider causes beyond the physical design and operation of the workplace and should widen their investigation to include the organization of work roles and the autonomy of the workforce."

A combination of measures can help reduce the likelihood your office will experience sick building syndrome, including increasing ventilation and air distribution, removing known pollutants, replacing water-stained ceiling tiles and carpets, introducing air filtration, and educating management and maintenance personnel. Heating, ventilation, and air-conditioning systems should, at a minimum, meet local building code ventilation standards. And time should always be allowed for the off-gassing of chemical contaminants in new building materials before occupancy.

CONTACTS: EPA "Sick Building Syndrome" page, http://epa.gov/iaq/pubs/sbs.html.

Some people argue that recycling uses more energy than it saves, and thus it is not worth the effort. Is this true?

Tigger Fox, Millinocket, ME

Controversy over the benefits of recycling bubbled up in 1996 when columnist John Tierney posited in a *New York Times Magazine* article that "recycling is garbage." He wrote, "Mandatory recycling programs . . . offer mainly short-term benefits to a few groups—politicians, public relations consultants, environmental organizations, and waste handling corporations—while diverting money from genuine social and environmental problems. Recycling may be the most wasteful activity in modern America."

Environmental groups were quick to dispute Tierney's claims, especially assertions that recycling was doubling energy consumption and pollution while costing taxpayers more money than the disposal of plain old garbage. The Natural Resources Defense Council and the Environmental Defense Fund each issued reports detailing how municipal recycling programs reduce pollution and the use of virgin resources while decreasing the sheer amount of garbage and the need for landfill space—all for less, not more, than the cost of regular garbage pickup and disposal.

Michael Shapiro, then director of the EPA's Office of Solid Waste, also weighed in: "A well-run curbside recycling program can cost anywhere from $50 to more than $150 per ton . . . trash collection and disposal programs, on the other hand, cost anywhere from $70 to more than $200 per ton. This demonstrates that, while there's still room for improvements, recycling can be cost-effective."

But in 2002, New York City, a municipal recycling pioneer, found that its much-lauded program was losing money, so it eliminated glass and plastic recycling. According to the mayor, Michael Bloomberg, recycling plastic and glass was costing twice as much as disposal. Meanwhile, low demand for the materials meant that much of it was ending up in landfills anyway, despite best intentions.

Other major cities watched closely to see how New York was faring with its scaled-back program (the city never discontinued paper recycling). But then New York City closed its last landfill, and private out-of-state operators raised prices due to the increased workload of hauling away and disposing New York's trash. As a result, glass and plastic recycling became economically viable for the city again, and New York reinstated the program, with a more efficient system and with more reputable service providers than it had used previously.

According to syndicated "Straight Dope" columnist Cecil Adams, the lessons learned by New York are applicable everywhere. "Some early curbside recycling programs . . . waste resources due to bureaucratic overhead and duplicate trash pickups (for garbage and then again for recyclables). But the situation has improved as cities have gained experience." Adams also says that, if managed correctly, recycling programs should cost cities (and taxpayers) *less* than garbage disposal for any given equivalent amount of material.

Even though the benefits of recycling over disposal are many, keep in mind that it better serves the environment to "reduce and reuse" before recycling even becomes an option.

CONTACT: Natural Resources Defense Council, www.nrdc.org/cities/recycling/gnyc.asp.

What is light pollution, and is it really a factor in breast cancer?

Gudrun Smythe, Madison, WI

The glow of city lights blotting out stars in the night sky has frustrated many a stargazer, but recent studies have shown that light pollution (excess or obtrusive light at night) can actually have serious health effects. Researchers have found that exposure to bright nocturnal light can decrease the human body's production of melatonin, a hormone secreted at night that regulates our sleep-wake cycles. And decreased melatonin production has in turn been linked to higher rates of breast cancer in women.

"Light at night is now clearly a risk factor for breast cancer," says David Blask, a researcher at the Bassett Research Institute in Cooperstown, New York. "Breast tumors are awake during the day, and melatonin puts them to sleep at night," he adds.

Epidemiologist Richard Stevens of the Department of Energy's Pacific Northwest National Laboratory first discovered the link between breast cancer and light pollution in the late 1980s. Stevens found that breast cancer rates were significantly higher in industrialized countries (where nighttime lighting is prevalent) than in developing regions.

William Hrushesky of the South Carolina–based Dorn Veterans Affairs Medical Center discovered that female night-shift workers have a 50 percent greater risk of developing breast cancer than other working women. He also found that blind women have high melatonin concentrations and unusually low rates of breast cancer.

To reduce breast cancer risks from light pollution, *Prevention* magazine recommends nine hours of sleep nightly in a dark room devoid of both interior (computer screen) and exterior (streetlamp) light sources.

A study of twelve thousand Finnish women found that those who slept nine hours nightly had less than one-third the risk of developing a breast tumor than those who slept only seven or eight hours. Even bright light from a trip to the bathroom can have an effect, so dim night-lights are recommended for night lighting.

Light pollution causes other problems besides increased cancer risks. According to the Sierra Club, birds and animals can be confused by artificial lighting, leading them away from familiar foraging areas and disrupting their breeding cycles. And the photosynthetic cycles of deciduous trees (those that shed their leaves in the fall) have been shown to be disrupted due to a preponderance of artificial nighttime lighting.

Another environmental impact of excessive use of artificial light is, of course, energy waste. The International Dark-Sky Association computes that unnecessary nighttime lighting wastes upward of $1.5 billion in electricity costs around the world each year, while accounting for the release of more than twelve million tons of carbon dioxide into the atmosphere. Individuals can do their part by keeping lights dim or off at home at night and convincing their employers and local government offices to do the same.

CONTACTS: International Dark-Sky Association, www.darksky.org; Sierra Club, http://newyork.sierraclub.org/longisland/lightpollution.html.

I'm moving into a freshly painted apartment and am curious to know whether it makes any sense to repaint the walls with nontoxic paint in hopes of "covering up" the toxic stuff already there.

Erin East, New York, NY

You can breathe relatively easy unless your apartment was painted very recently. Conventional indoor paints do indeed release potentially

toxic chemicals during and shortly after application, but once the paint's dry the majority of the dangerous volatile organic compounds (VOCs) stay sealed up. So most people will not be affected once the telltale new paint smell has faded away.

If someone is suffering adverse health effects from exposure to fresh paint it should not be taken lightly. Off-gassing VOCs can cause serious respiratory tract irritation as well as visual impairment, headaches, dizziness, and memory loss. Additionally, many VOCs have been shown to cause cancer in animals, and some are suspected of being carcinogenic to humans. Health effects vary greatly depending on the particular chemicals involved, the amount of exposure, and the sensitivity of those living with them. Besides paints, a wide range of other home products—including building materials, carpets, furniture, cleaning supplies, and bug sprays—can emit VOCs.

If someone is suffering from respiratory problems or other symptoms upon moving into a freshly painted residence, or remaining sensitive long after a paint job, there are many paints now on the market that can help decrease the amount of VOCs emitted into the air.

There are essentially three general categories of nontoxic (or low-toxic) paints: zero-VOC, low-VOC, and so-called natural. Keep in mind, however, that the term "nontoxic" is used in its broadest sense. Even zero-VOC formulations, such as those made by AFM Safecoat, YOLO Colorhouse, and Ecos, for example, can contain trace amounts (up to five grams per liter or less) of toxic ingredients.

Some leading low-VOC paint manufacturers are Cloverdale, Vista, and Miller. Industry leaders Benjamin Moore and Sherwin-Williams also offer their own low-VOC lines. "Natural" paints and finishes, from manufacturers such as Livos, Aglaia, and BioShield, are made from raw ingredients such as water, plant oils, clay, and milk protein, and usually contain minimal amounts of VOCs. Consumers can track down such healthier paints at retailers like ecohaus (formerly Environmental Home Center), Green Home, and even at some of the larger home-repair chains.

Precautions should be taken during the application of any paint. Only buy exactly what you need, and apply it with adequate ventilation. Remember to always keep paints out of the reach of children and pets, and safely dispose of all unused product. If ventilation is not sufficient, wear a respirator with a filter that will capture and prevent the inhalation of VOCs.

CONTACTS: ecohaus, www.environmentalhomecenter.com; EPA "Introduction to Indoor Air Quality: Volatile Organic Compounds," http://epa.gov/iaq/voc.html; Green Home, www.greenhome.com.

Are the flame retardants used in many products as a fire-safety precaution dangerous to our health? If so, what can I do to avoid contact with them?

Katya, via e-mail

Flame retardants are in widespread use in both the United States and Canada, primarily in carpet padding, foam cushions, polyester bedding, clothing, wallpaper, and the plastic housings for computers, fax machines, and other electronics. Most are made from variations of a chemical known as PBDE, which stands for polybrominated diphenyl ether.

In laboratory studies, some prenatal exposure to PBDEs caused permanent problems in rodent brain development. Scientists are quick to point out, though, that levels in humans have not (yet) reached the levels that cause problems in lab animals, but scientists are concerned because the levels in humans keep rising.

PBDEs are "persistent" in that they don't break down but remain active in our air, water, soil, and food. Washington State researchers say that PBDEs are building up in animals throughout the food chain, even turning up in killer whales in Washington's Puget Sound and in the bodies of polar bears in the Arctic.

PBDEs also stay in our bodies, accumulating in our fatty tissue. The United States is the world's largest maker and user of PBDEs, and levels found in Americans are as much as one hundred times higher than those found in people in Europe, where most PBDEs were banned in 2001. North American levels, say scientists, are doubling every two to five years. Primarily, human exposure has been through eating fish, though babies can be exposed by drinking mother's milk. Children are also exposed when they wear polyester pajamas treated with flame retardants. Indeed, PBDE chemicals easily "off-gas" from the very products they are designed to make safe.

Consumers can take precautions and avoid products that contain PBDE. Among other cautions, the Healthy Children Project recommends buying clothing, bedding, and furniture made from natural fibers, such as cotton and wool, which do not melt near heat and do not need to contain flame retardants.

Another way to minimize exposure is to stick to a diet low in animal fat, since the chemicals accumulate in larger amounts in animals higher up the food chain. Joyce Newman of the *Green Guide* recommends vegetables, fruits, and whole grains over meat and fish. When choosing meat and fish, she suggests cutting away as much of the fat as possible and choosing leaner cuts overall.

As for consumer goods, researchers says that industries need to rethink their product designs, avoiding highly flammable materials and keeping ignitable materials separated or shielded from heat sources. Some mattresses in use now in nursing homes and hospitals employ a "barrier layer" of durable material between the surface fabric and interior foam and meet stringent fire-safety standards without the use of chemicals.

CONTACTS: Healthy Children Project, www.healthychildrenproject.org; the *Green Guide*, www.thegreenguide.com.

How is it that African Americans are said to suffer the most in the United States from pollution and other environmental ills?

Jon Stein, Novato, CA

In 1979, while conducting postdoctoral sociology research in Houston, Dr. Robert Bullard noticed that all the city's garbage dumps were located in and around African American neighborhoods, even though blacks only accounted for a quarter of the city's population. Bullard hypothesized that such discriminatory siting was no coincidence, especially since Houston had no zoning laws to regulate land use. At the time, his findings helped a middle-class African American community in the city prevent the building of a new dump facility in their neighborhood.

Bullard cast his net wider to find more examples of what he called "environmental racism." He found not only dumps but also polluting factories and other industrial blemishes throughout the American Southeast—from West Virginia to Alabama and Texas and Louisiana and Florida—located where poor and sometimes middle-class African Americans lived. While discriminatory decision making was no doubt a factor, Bullard also theorized that such communities' lack of political experience also contributed to their predicament. Such realizations gave birth to a new political movement, and today thousands of activists in the United States and elsewhere monitor policy making, lobby for new laws, and fight city hall in the struggle for "environmental justice."

Dr. Bullard says that the kinds of problems he uncovered in black communities in the Southeast are not limited to a particular region or ethnicity. "People of color in all regions of the country bear a disproportionate share of the nation's environmental problems," he says in his seminal book *Dumping in Dixie*.

Bullard's pioneering work also helped shatter the myth that minority communities didn't care about the environment. With financial help from the Charles Stewart Mott Foundation, Bullard convened the First National People of Color Environmental Leadership Summit held in October 1991 and a year later published the first version of the *People of Color Environmental Groups Directory* with listings for more than three hundred different groups in the United States alone. An expanded version of the directory, released in 2000, is available free online from the website of Bullard's Environmental Justice Resource Center at Clark Atlanta University.

These days, Bullard is marshalling all the resources he can to monitor the "mother of all cleanups" in post-Katrina New Orleans and has been highly critical of the slow pace of federal and state efforts. Acknowledging that funds are limited, Bullard wonders "which neighborhoods will get cleaned up and which ones will be left contaminated."

CONTACTS: Environmental Justice Resource Center at Clark Atlanta University, www.ejrc.cau.edu; *People of Color Environmental Groups Directory,* www.ejrc.cau.edu/poc2000.htm.

Is it true that the materials used in car interiors can be hazardous to our health?

Chris Smith, Bethesda, MD

The health hazards lurking inside car interiors—occupied by Americans an average of ninety minutes each day—have largely escaped the intense scrutiny given recently to indoor air pollution.

Fortunately, the Michigan-based Ecology Center has now detailed how heat and ultraviolet (UV) light inside cars can trigger the release of a number of chemicals linked to birth defects, premature births, impaired learning, and liver toxicity, among other serious health problems.

Polybrominated diphenyl ethers (or PBDEs, often used as fire retardants) and phthalates (chemicals used to soften plastics) are the primary culprits. Part of the seat cushions, armrests, floor coverings, and plastic parts in most car interiors, these chemicals are easily inhaled or ingested by drivers and passengers through contact with dust. The risks are greatest in summer, when car interiors can get as hot as 192 degrees Fahrenheit.

Motorists can lessen their risks by rolling down car windows, parking in the shade, and using interior sun reflectors. But the Ecology Center is urging carmakers to stop using such chemicals in the first place. "We can no longer rely just on seat belts and air bags to keep us safe in cars," says Jeff Gearhart, the Ecology Center's clean car campaign director and coauthor of its report on the subject. "Our research shows that autos are chemical reactors, releasing toxins before we even turn on the ignition."

In preparing its report, the Ecology Center collected windshield film and dust from 2000 to 2005 models made by eleven leading manufacturers. Volvo was found to have the lowest phthalate levels and the second-lowest PBDE levels, making it the industry leader in interior air quality. Volvo also has the toughest policies for phasing out these chemicals. Other makers claim they have eliminated some but not all PBDEs and phthalates. Ford, for example, reports that it has eliminated PBDEs from "interior components that customers may come into contact with." Honda reports it has eliminated most phthalate-containing PVC. Other carmakers tested were BMW, Chrysler, GM, Hyundai, Mercedes, Subaru, Toyota, and Volkswagen.

Indoor air pollution is listed by the EPA as one of the top five environmental risks to public health, but the Ecology Center is especially concerned that concentrations of PBDEs are five times higher inside cars than in homes and offices. The organization is calling on the U.S. government to ban the worst forms of PBDEs and phthalates from use in any indoor environments and has enlisted the help of several concerned members of Congress to help write legislation to that effect.

CONTACT: Ecology Center, www.ecocenter.org.

Where can I find carpeting without the strong odors and health concerns of conventional synthetic materials?

Denise Purdy, via e-mail

Wall-to-wall carpeting, once touted as the ultimate in luxury, is now getting a decidedly skeptical second look from the chemically sensitive. But carpets made from all-natural materials are now readily available from companies such as Earth Weave and Natural Home. Their attractive carpets are entirely biodegradable and are made of wool, jute, hemp, and rubber. Both companies pride themselves on making products free of toxic dyes and mothproofing or stain-repellant chemicals.

These carpets are becoming more popular in part because there are, on average, 120 chemicals in each new piece of conventional synthetic carpeting, including the adhesive. Many of these chemicals are suspected or known carcinogens, such as formaldehyde. According to a spokesperson for Antibody Assay Laboratories, which provides services to health care providers, "These chemicals 'off-gas' into the environment, polluting indoor air with volatile organic compounds that can create symptoms from itchy eyes to shortness of breath, headaches, and nausea."

If you must install synthetic carpet, make sure you air it out before putting it in place. And consider using less-toxic installation techniques, such as that developed by TacFast Systems International in Ontario, Canada—a hoop-and-loop method similar to Velcro that eliminates the need for liquid adhesives. Another environmentally conscious backing choice is all-natural wool carpet padding from the Seattle-based ecohaus (formerly Environmental Home Center), which ships worldwide. The padding is made from a variety of wool fibers, without dyes or fire retardants, and is mechanically needled (not glued) to a biodegradable jute backing.

One additional way to live with traditional carpeting is to fill the room with plants that have been shown to absorb toxins, including aloes, philodendrons, and spider plants.

CONTACTS: AFM Enterprises, (619) 239-0321, www.afmsafecoat.com; Earth Weave, (706) 278-8200, www.earthweave.com; ecohaus, www.environmentalhomecenter.com, (800) 281-9785; Natural Home, (707) 571-1229, www.naturalhomeproducts.com; TacFast Systems, (905) 886-0785, www.tacfastsystems.com.

What's the story with electromagnetic fields? Can you really get cancer from living near clusters of power lines or from sleeping near the fuse box in your house?

Tim Hutchins, Arcata, CA

Over the past twenty-five years, there has been growing concern and controversy, both among scientists and the public, about possible links between electromagnetic fields (EMFs) and any of several forms of cancer.

EMFs are invisible lines of force that radiate from sources of electricity, including power lines and transformers, interior home wiring, and all electrical appliances, gadgets, and machinery. These fields have both electric and magnetic components that diminish in strength with distance. The electric segment of the field may be at least partially blocked by physical barriers such as walls, trees, and partitions, but the magnetic segment is much less easily shielded.

The federally funded National Institute of Environmental Health Sciences (NIEHS) concluded in 1998 that there was still no clear answer to the question of risk; it did affirm that extremely low-frequency (ELF) EMFs should be classified as possible human carcinogens in the case of two cancers: childhood leukemia related to residential exposure; and chronic lymphocytic leukemia in adults in occupational settings.

The World Health Organization concluded around the same time, based on studies of childhood leukemia, that ELF magnetic (but not electric) fields were possibly carcinogenic to humans.

But uncertainty remains. One of NIEHS's key conclusions in 1998 was: "Despite a multitude of studies, there remains considerable debate over what . . . health effects result from exposure to EMF. There is still no clear answer to the question, 'Can exposure to electric and magnetic fields resulting from production, distribution, and use of electricity promote cancer or initiate other health problems?'" NIEHS decided there was inadequate evidence to draw any clear conclusions.

But May Dooley, whose company Enviro Health Environmental Home Inspections provides comprehensive on-site EMF testing, cites several scientific studies showing that EMF exposure has increased the size and number of tumors in laboratory animals. She recommends reducing exposure as much as possible: "If someone with cancer knew that eating a certain food would speed up the growth of tumors, you can bet that he or she wouldn't eat that food."

CONTACTS: Enviro Health Environmental Home Inspections, (888) 735-9649, www.create-your-healthy-home.com; National Institute of Environmental Health Sciences, www.niehs.nih.gov; World Health Organization, www.who.int/peh-emf/en.

How can I reduce the amount of unwanted mail that I receive?

Jennifer Pearle, Brattleboro, VT

The Center for a New American Dream (CNAD) says some 5.6 million tons of catalogs and other direct-mail solicitations clog U.S. landfills every year. Meanwhile, the average household may receive as many as one thousand unwanted pieces of mail annually, of which only about 22 percent ever gets recycled. Further, says CNAD, each American will spend about eight *months* of his or her life opening unwanted mail! So reducing the volume of mail you receive will not only saves trees, energy, and landfill space, it will also save you time.

Registering with the Mail Preference Service of the Direct Marketing Association (DMA) can help reduce the mountain of junk mail you receive, but it is no guarantee. DMA includes your name in a "do not mail" database, but marketers aren't obligated to consult it. In fact, most of the mail you receive comes from companies that rented your name from somewhere else. Fortunately, most large mailers do use the service routinely, because they know that there's nothing to gain by mailing to people who don't want mail.

Another way to lighten your mailbox is to go to optoutprescreen.com, where you can get yourself off lists that credit card and insurance companies use to solicit your business. It's a centralized website run by the country's four major credit bureaus: Equifax, Experian, Innovis, and TransUnion. These are the companies that businesses check with before accepting your credit card. They are also the largest sources of names and addresses used by credit card companies to trawl for new customers via mass mailings. Luckily, the federal Fair Credit Reporting Act requires that these bureaus delete any person's name and address from rented lists if they so request.

You should also notify (in writing) all of the companies you do business with (especially credit card brokers, magazine publishers, and catalog merchants) that you want to be in their "do not promote" or "in-house suppress" file. It's most effective if you give them the message the very first time you make a transaction, but it can be done at any time.

Track corporate responsibility by cleverly altering your name. John Smith, for example, can subscribe to *Rolling Stone* magazine as John R. S. Smith and ask that his name be left off any lists. If he later receives mail from another entity addressed to John R. S. Smith, he'll know precisely how they found him and can take action accordingly.

The website Junkbusters provides further guidelines for reducing mail and other intrusive marketing.

CONTACTS: Direct Marketing Association's Mail Preference Service, www.dmaconsumers.org/consumerassistance.html; Junkbusters, www.junkbusters.com; optoutprescreen.com, www.optoutprescreen.com.

Is it bad for the environment to dump clog removers like Drano down the drain? What are some alternatives to such products?

Cindy Jones, via e-mail

The active ingredient in Drano and other conventional drain cleaners is sodium hydroxide, otherwise known as caustic soda or lye. It is a man-made chemical used for its corrosive properties. According to the federal Agency for Toxic Substances and Disease Registry, lye is not a pollutant per se, as it separates into relatively harmless component elements once released into water or moist soil.

But sodium hydroxide is an irritant that can burn skin and aggravate nose, throat, and respiratory airways, so contact with it is best avoided. If ingested outright, it will likely induce vomiting as well as cause chest or abdominal pain and make swallowing difficult—so keep it well out of the reach of children. Even smaller kids might be able to find it if, like most people, you store it under the sink.

Safer alternatives do exist. A simple plunger or mechanical drain snake—along with a little elbow grease—can often free up clogs as well as or better than sodium hydroxide compounds. One home remedy with a proven track record is to pour a handful of baking soda mixed with a half cup of vinegar down the drain, and follow it quickly with boiling water.

Another option is to use one of the many enzymatic biological drain cleaners on the market today, such as Earth Friendly Products' Natural Earth Enzyme Drain Opener or Bi-O-Kleen's Bac-Out. These use natural bacterial and enzyme mixtures to open and keep drains clear. And unlike sodium hydroxide, they are noncaustic and will not cause a fire.

As any plumber will tell you, a good maintenance regimen is the best way to prevent clogged drains. Flushing drains weekly with boiling water can help keep them clear. Also, installing small screens atop drains will help keep hair, lint, and other clogging elements out of the pipeline in the first place.

CONTACTS: Agency for Toxic Substances and Disease Registry, www.atsdr.cdc.gov; Bi-O-Kleen, www.bi-o-kleen.com; Earth Friendly Products, www.ecos.com.

Is the chlorine bleach used for whitening clothes bad for the environment? And if so, what are some safe alternatives?

Nancy Potter, via e-mail

More than 80 percent of American households use chlorine bleach to whiten their clothes and clean inside their homes, but most consumers don't realize that the use of this seemingly innocuous cleaning additive could be polluting their home as well as the great outdoors.

"The fumes of cleaners contain a high concentration of chlorine that when breathed in can irritate the lungs and be particularly dangerous for people who suffer from heart conditions or chronic respiratory problems such as asthma or emphysema," says Patty Avey, editor of *Smart Living News.* "When the fumes are emitted in small, poorly ventilated rooms such as the bathroom, the risks are increased," she adds.

Another immediate risk of having chlorine bleach around the house is accidental ingestion by little ones—poison-control centers across the country receive about twenty thousand such calls each year. Also, combining chlorine bleach with ammonia and other acids can cause deadly fumes.

Though the EPA says there's no danger in using chlorine bleach around the house because the amount of chlorine is too low to warrant serious concern, the agency does regulate its use for industrial purposes and confirms links between exposed workers and lung irritation.

At home or in the factory, chlorine is a big problem for the environment once it is discarded or rinsed away. It bonds with other chemicals in the wastewater stream to form carcinogenic organochlorines (such as dioxin) that contaminate drinking water supplies, among other risks.

But healthy and environmentally safe alternatives to chlorine bleach abound. Many of these can be made at home with household products you probably already have. Half a cup of hydrogen peroxide can work well as a bleach alternative when diluted with warm water prior to going in the wash load.

If you're not that ambitious, commercial variations on such formulas, which give consumers the benefit of oxygen-based stabilizers that ensure even distribution within wash loads, are available from companies such as Seventh Generation, Earth Friendly Products, and Bio Pac. Most of these products are available at natural food stores as well as online and at larger, well-stocked supermarkets.

But before spending a fortune on bleach alternatives, consumers should see if hard water might be causing their clothes to look gray and dingy from soap scum and mineral deposit buildup. Clues that you might have hard water include clean dishes with water spots on them, white and crusty sediment on fixtures, or a recurring bathtub ring. If you do have hard water, simply add enough baking soda to the laundry to make the wash water feel slippery to the touch and see if that doesn't whiten whites and brighten colors.

CONTACTS: Bio Pac, www.bio-pac.com; Earth Friendly Products, www.ecos.com; EPA chlorine fact sheet, www.epa.gov/chemfact/f_chlori.txt; Seventh Generation, www.seventhgeneration.com; *Smart Living News*, www.smartlivingnews.com.

I'm sick of maintaining my lawn, and I'm sure that all the chemicals I'm using are no good for the environment. What alternatives can I explore that will save time and money while keeping the property looking nice?

Sarah, Bethesda, MD

Grass lawns first appeared in Europe in medieval times and were status symbols for the rich. In those days, they were trimmed by the fairly simple expedient of grazing livestock on them; polluting lawn mowers and poisonous weed killers hadn't been invented yet. Lawns did not become popular in North America until the middle of the twentieth century but are now as common as the middle-class suburban homes they surround.

Besides hogging public water supplies—over 50 percent of U.S. residential water usage goes to irrigating lawns—a 2002 Harris survey

found that American households spend twelve hundred dollars per year on residential lawn care. Indeed, the booming lawn-care industry is more than eager to convince us that our grass can be greener—and then sell us all the synthetic fertilizers, toxic pesticides, and leaky lawn mowers to make it so.

According to the website Eartheasy, there are many alternatives to a carpet of monochromatic grass for one's property. The site recommends ground-cover plants and clover, which grow horizontally, low to the ground, and require no cutting. Some varieties of ground cover are alyssum, bishop's-weed, and juniper. Common clovers include yellow blossom, red clover, and Dutch or white clover, the best suited of the three for lawn use. Ground cover plants and clovers naturally fight weeds, act as mulch, and add beneficial nitrogen to the soil.

Eartheasy also recommends flower and shrub beds, which can be "strategically located to add color and interest while expanding the low-maintenance areas of your yard," and planting ornamental grasses. Ornamental grasses, many of which flower, have numerous benefits over conventional grasses, including low maintenance, little need for fertilizer, minimal pest and disease problems, and resistance to drought.

David Beaulieu says in About.com's "Guide to Landscaping" that moss plants should also be considered, especially if your yard is shady: "Because they are low-growing and can form dense mats, moss plants can be considered an alternative ground cover for landscaping and planted as 'shade gardens' in lieu of traditional lawns." Moss plants do not possess true roots, he points out, instead deriving their nutrients and moisture from the air. As such, they like wet surroundings and also soil with a pH that is acidic.

Of course, lawns aren't all bad. They make great recreational spaces, prevent soil erosion, filter contaminants from rainwater, and absorb many kinds of airborne pollutants. So even if you go in other directions, you might still keep a small section of lawn, which can be mowed with a few easy strokes. If you do, the EPA recommends avoiding traditional synthetic fertilizers, herbicides, and pesticides. A number of all-natural alternatives are now widely available at nurseries. Natural lawn-care advocates also advise mowing high and often so that grass can outcompete any nascent weeds. Also, leaving clippings where they land–so they can serve as natural mulch–helps prevent weeds from getting a foothold.

CONTACTS: Eartheasy, www.eartheasy.com/grow_lawn_alternatives.htm; About.com, http://landscaping.about.com/cs/groundcovervines1/p/moss.htm; EPA publication *Healthy Lawn, Healthy Environment*, www.newdream.org/consumer/lawncare.pdf.

I've heard that gas-powered lawn mowers, despite their small engine size, actually pollute as much as cars. If this is true, is there a greener way to cut my grass?

Jon Haufe, Seattle, WA

Lawn mowers are indeed big polluters. A Swedish study conducted in 2001 concluded, "Air pollution from cutting grass for an hour with a gasoline-powered lawn mower is about the same as that from a 100-mile automobile ride." Meanwhile, the fifty-four million Americans mowing their lawns each weekend with gas-powered mowers may be contributing as much as 5 percent of the nation's air pollution, says the EPA.

The problem is that small engines emit disproportionately large amounts of carbon monoxide, volatile organic compounds, and nitrogen oxides that contribute to smog. The human health effects of smog-laden air are well known and include inflammation and damage to lungs, increased risk of asthma attacks, and lowered levels of oxygen in the bloodstream, which can aggravate heart conditions.

The good news is that the EPA has phased in new emissions standards for gas-powered mower engines, resulting in a 32 percent smog reduction for mowers made in 2007 or later. And California has even more stringent standards.

But gas power is not the only option: you can also go electric. The easy part is the price, as many models cost less than two hundred dollars. The trade-off for plug-in mowers is that they only work for small lawns and must be tethered to a power outlet during use. Also, going electric is not necessarily a way to reduce pollution overall. According to *Consumer Reports*, "Achieving a net environmental savings from switching to electric mowers depends on the efficiency of the power plant" where the electricity originates.

If money is not an issue, the twenty-five-hundred-dollar solar-powered "auto mower" from Husqvarna can't be beat for both eco-friendliness and convenience. It wanders unattended around any level lawn, its collision sensors carefully avoiding contact with anything but the grass itself. While it is currently not available directly in the United States, some Husqvarna dealers are willing to special-order it from Sweden where it is manufactured.

Of course, the greenest choice of all is the mower that runs on three square meals a day and a good exercise regimen: the venerable human-powered reel mower. The most popular choices are from American Lawn Mower, which makes nine models including a child-size one. They can be found at retailers like Ace Hardware and Target (and at local hardware stores) and in catalogs like Real Goods and Smith & Hawken.

CONTACTS: American Lawn Mower, www.reelin.com; Gaiam, www.gaiam.com; Husqvarna, www.husqvarna.com.

What kinds of home improvements could I do that would make my house healthier and more environmentally friendly?

Elizabeth Bram, via e-mail

Indoor air quality, as much as five times worse than outdoor air, is probably the first frontier. According to Glenn Haege, a master handyman who hosts a national radio show on home repair, as our homes and apartments have become more energy efficient and airtight, "humidity levels from cooking and breathing tend to increase, causing mold and mildew." Harmful chemicals, he says, from construction materials, insulation, furniture, carpeting, padding, paints, solvents, and household cleaners, drawn by this moist atmosphere, combine to contaminate the indoor air, which then stays trapped inside.

The first step in remedying this problem is to test your indoor air. Pure Air and EnviroLogix, among others, sell inexpensive and easy-to-use indoor-air-quality testing kits. Once you get an idea of what contaminants are floating around your home, you can get to work fixing the problem. Green superstores such as ecohaus, Green Building Supply, and Oikos offer greener and healthier building supplies and materials. Also, BuildingGreen offers a free online "GreenSpec" database at its website, with detailed listings for over two thousand environmentally preferable building products.

Materials *outside* the home can also contribute to health problems. One example is pressure-treated lumber, which contains a form of cyanide to keep pests away. Kids who play on backyard jungle gyms and decks made of such material can develop rashes and skin infections. Cedar wood is a naturally pest-resistant alternative that, while more expensive, is a kinder and gentler option that will stand the test of time.

Other ways to green up the home include replacing traditional incandescent lightbulbs with more energy-efficient compact fluorescents as well as switching out conventional hot-water heaters in favor of solar or on-demand tankless versions. And for saving on water, replacing traditional showerheads and toilets with pressurized low-flow alternatives can save gallons per day while generating cost savings on utility bills. Likewise, capturing rainwater and shower "gray water" to irrigate the garden is another smart move.

Do-it-yourselfers can find hundreds of websites offering tips on green building and repair. Glenn Haege's masterhandyman.com and naturalhandyman.com both offer a plethora of articles and links and are good resources if you're looking to improve your own handy skills while staying true to your green ideals. Two helpful books are *Green Remodeling* by David Johnston and Kim Master and *Green Building Materials: A Guide to Product Selection and Specification* by Ross Spiegel and Dru Meadows. For less-handy home owners, finding a handyman well versed in green building issues might be a better way to go. The Natural Handyman Network offers a free online search tool that should offer some promising leads.

CONTACTS: BuildingGreen, www.buildinggreen.com; Enviro-Logix, www.envirologix.com; ecohaus, www.environmentalhomecenter.com; Green Building Supply, www.greenbuildingsupply.com; Master Handyman, www.masterhandyman.com; Natural Handyman Network, www.naturalhandyman.com; Oikos, www.oikos.com.

Do houseplants really help to clean indoor air?

Jackson Schlemmer, London, England

Great claims are being made for the humble houseplant. NASA researcher Bill Wolverton first reported in 1984 that some common houseplants cleaned polluted indoor air. He found that philodendrons and golden pothos excelled at stripping formaldehyde from the air, gerbera daisies and chrysanthemums wiped out excessive amounts of indoor benzene, and pot mums and peace lilies absorbed a toxic degreasing solvent known as TCE.

A later NASA study, also conducted by Wolverton, saw houseplants removing up to 87 percent of toxic indoor air within twenty-four hours. And a 1994 German study reported that one spider plant could cleanse a small room of formaldehyde in just six hours. Further, English ivy, bamboo palm, and snake plants have been shown to be effective in removing cigarette smoke as well as noxious odors from carpeting and chemical-laden household cleaners.

There were and are some skeptics, of course. Hal Levin, a research architect in Santa Cruz, California, and editor of *The Indoor Air Bulletin*, said at the time that Wolverton's studies were "interesting," but he told the *New York Times* that "the notion that plants remove pollutants from indoor air has not been adequately proven. To build a building using this technology, given the state of knowledge, is just not justified."

So how does the houseplant supposedly help clean the air? The answer lies in its basic ability to absorb carbon dioxide from the air while releasing oxygen as part of the photosynthetic process. Houseplants are essentially doing indoors what other plants and trees ordinarily do outdoors.

To maximize the benefits of houseplants in cleaning indoor air, it is generally recommended to use one plant for every hundred square feet of indoor space. Besides those plants mentioned above, other good indoor air cleaners include palms, ferns, dracaenas, corn plants, weeping fig, dumb cane, orchids, arrowheads, dwarf banana, and Chinese evergreen.

The EPA recommends opening the windows and letting in some good old-fashioned fresh air as the best antidote to breathing in "off-gassed" airborne toxins in both homes and offices. But many modern buildings do not permit such exchanges between indoor and outdoor air, and it is in just these situations where houseplants can really make the difference.

CONTACT: Plant-Care.com; www.plant-care.com/indoor-plants-clean-air-1.html.

Is it more environmentally friendly to hand wash dishes or use a dishwasher?

Jennifer Furnari, Sonora, CA

Dishwashers are the way to go if you comply with two simple criteria: "Run a dishwasher only when it's full, and don't rinse your dishes before putting them in the dishwasher." So says John Morril of the American Council for an Energy Efficient Economy (ACEEE), who also advises not using the dry cycle. The water used in most dishwashers is hot enough, he says, to evaporate quickly if the door is left open after the wash and rinse cycles are complete.

Scientists who studied the issue found that the dishwasher uses only half the energy, one-sixth of the water, and less soap than hand washing an identical set of dirty dishes. Even the most sparing and careful washers could not beat the modern dishwasher. The study also found that dishwashers excelled in cleanliness over hand washing.

Most dishwashers manufactured since 1994 use seven to ten gallons of water per cycle, while older machines use eight to fifteen gallons. Newer designs have improved dishwasher efficiency immensely. Hot water can now be heated in the dishwasher itself, not in the household hot-water heater, where heat gets lost in transit. Dishwashers also heat only as much water as needed. A standard twenty-four-inch-wide household dishwasher is designed to hold eight place settings, but some newer models will wash the same amount of dishes inside an eighteen-inch frame, using less water in the process. If you have an older, less-efficient machine, ACEEE recommends hand washing for the smaller jobs and saving the dishwasher for the dinner party's aftermath.

New dishwashers that meet strict energy and water-saving efficiency standards can qualify for an Energy Star label from the EPA. Besides being more efficient and getting the dishes cleaner, qualifying newer models will save the average household about twenty-five dollars per year in energy costs.

Like John Morril, the EPA recommends always running your dishwasher with a full load and avoiding the inefficient heat-dry, rinse-hold, and prerinse features found on many recent models. Most of the energy the appliance uses goes to heat the water, and most models use just as much water for smaller loads as for larger ones. Propping the door open after the final rinse will dry the dishes when the washing is done.

CONTACTS: American Council for an Energy Efficient Economy, www.aceee.org/consumerguide/topdish.htm; Energy Star, www.energystar.gov/index.cfm?c=dishwash.pr_dishwashers.

Are there any environmental or health drawbacks to putting vinyl siding on my house?

Charles Leach, via e-mail

Vinyl siding may not pose identifiable risks once installed properly on your home, but its production and disposal contribute to a wide range of health and environmental problems. In producing polyvinyl chloride (PVC), the basic element in all vinyl products, workers are exposed to a multitude of hazardous chemicals. These include chlorine gas, which can cause eye and skin irritation and breathing difficulties in the short term and lung disease, among other maladies, from prolonged exposure.

Meanwhile, according to Greenpeace, the production process releases other dangerous chemicals, such as dioxin, into the environment surrounding PVC factories. Dioxin nearly wiped out the bald eagle in the lower forty-eight states, and it has been linked to cancer, endometriosis, neurological damage, immune system damage, respiratory problems, liver and kidney damage, and birth defects in humans.

An even larger problem is that there is no responsible way to dispose of PVC and vinyl siding at the end of its life cycle. Landfills do not knowingly accept it, as it can pollute groundwater and result in dioxin-forming landfill fires. And unfortunately, vinyl cannot be recycled due to the chlorine used in its production. If mixed inadvertently into a recycling load, vinyl will contaminate everything therein.

Incinerating vinyl releases poisonous chlorine gas as well as dioxin into the air. When a house with vinyl siding catches fire, dioxin and other toxic gases escape into the air, posing an even greater threat than the fire itself in some cases. It is not unusual, firefighters say, for people trapped in building fires to die of exposure to chemically toxic fumes before the flames actually reach them. Recently, a vinyl scrap yard fire forced the evacuation of two hundred people from a Virginia community, while another created a major airborne dioxin hazard in Ontario.

"We know enough about the dangers of PVC to begin to phase it out," says Lois Gibbs, the founder of the Virginia-based Center for Health, Environment and Justice. "We need to tell corporations to protect our health and environment by switching to non-PVC materials." Gibbs is the housewife turned activist who spurred the government into creating its Superfund program to clean up contaminated waste sites around the country after she discovered in 1978 that her neighborhood in Love Canal, New York, was located on a twenty-thousand-ton chemical waste dump.

Safer alternatives to vinyl siding do exist. According to the organization Greenaction, siding made from wood, fiber-cement board, or polypropylene is better for the environment and for human health. While some of these materials are available at Home Depot, local stores selling only green building materials would offer the best selection.

CONTACT: Center for Health, Environment and Justice, www.besafenet.com; Greenaction, www.greenaction.org; Greenpeace, www.greenpeaceusa.org.

Are there more environmentally friendly ways to deice pavement than using salt?

Heidi David, Concord, NH

Although salt and various salt derivatives melt ice effectively and make both walkways and roads safer, they can be damaging to the environment. After salt is applied, it washes off paved surfaces into storm drains or onto adjacent ground and can then be carried into nearby bodies of water. This salty runoff poisons fish and wilts vegetation. It also corrodes metals, damages concrete, and poses health risks to people with high blood pressure.

Some studies have also shown that salt applied to road surfaces increases automobile collisions with wildlife, especially white-tailed deer that are attracted to natural and artificial salt deposits in their normal course of feeding.

Despite these facts, salt remains the cheapest and most effective way to keep pavement free of ice. According to materials consultant Henry Kirchner, individuals can effectively use salt with minimum impact: "Do not use a chemical deicer to melt every bit of ice," says Kirchner. "Use only enough to break the ice-pavement bond, then remove the remaining slush by plowing or shoveling." All snow should be cleared away first, and the ice should be chipped off and moved away from water supplies and vegetation.

For small jobs, it may be feasible to use more potent, less environmentally toxic deicers like magnesium chloride or calcium magnesium acetate instead of rock salt. These stronger, though more expensive, compounds can be strategically applied before a storm to block ice from forming. Sand and cat litter can be used to provide temporary traction, but these materials may clog surface water and bury plants. Although many researchers are experimenting with even more benign deicers, including by-products of corn and cheese processing, none of these compounds is currently available to consumers.

Perhaps the larger issue is how municipalities store and use large amounts of road salt. Many of the most severe cases of environmental contamination have been caused by improper storage. When salt is stored outside uncovered, rain and snow can carry large quantities to surrounding soil and water.

As to reducing salt use, many cities and towns simply don't deice in flat residential areas, except during ice storms. Some use a mix of sand and salt instead of pure salt. Also, salt-spreading equipment that is well maintained will distribute salt more accurately and, as a result, use less. Additionally, salt that is wet before being spread sticks better to the road.

According to the trade magazine *Better Roads*, a product called Verglimit, a mixture of deicing salts and caustic soda, can be mixed with asphalt roadway during paving. Its installation doubles the cost of surfacing a road but helps reduce the amount of salt needed for deicing roadways and, according to the magazine, "in certain conditions can eliminate the need for salting entirely."

CONTACTS: *Better Roads,* "Materials for Deicing and Anti-icing," http://obr.genpublishing.com/articles/newprods/apr03bid.htm; Salt Institute, (703) 549-4648, www.saltinstitute.org.

I believe that I developed asthma from the fiberglass insulation in our home. How can I find insulation that won't make me sick?

Cynthia Bacon, Orlando, FL

Fiberglass, a common home insulator that replaced dangerous asbestos, is itself now associated with a range of health issues. Microscopic slivers of fiberglass can break loose during handling and be inhaled, irritating the lining of the respiratory tract and becoming lodged in lung tissue. This can cause a fibrous buildup that reduces lung capacity, or cause DNA mutations that can lead to lung cancer. In fact, cancer warnings appear on all fiberglass insulation sold in the United States.

Although wearing a respirator or dust mask can prevent inhalation of fibers during installation, all three principal U.S. manufacturers of fiberglass insulation now seal their batts in a perforated polyethylene or polypropylene sheeting so as to prevent airborne exposure. Nevertheless, for those suffering from aggravated respiratory problems, replacing fiberglass insulation with a more environmentally friendly alternative may be the best option. There are many such options available.

An environmental favorite is cellulose, which is made from recycled, shredded newspaper. In his book *The Solar House*, author Dan Chiras calls cellulose "one of the most environmentally friendly insulation choices." It is also highly efficient, readily available, and economically priced, he says, and thus competes well with fiberglass.

Chiras also recommends cotton insulation, calling it "a natural product and safe from a human health standpoint," while acknowledging that it is twice the price of fiberglass and "one of the most chemically intensive crops grown in the U.S." It contains no formaldehyde binders, however, a health and environmental plus, and usually contains a fire retardant, an important safety consideration.

Radiant barriers are another option, says the Fiberglass Information Network. Ideal for hot climates, they are made from metal foil and either kraft paper or bubble wrap. The network also recommends insulation batts made from recycled number 1 plastic, known as PET, the same material used to make some soda bottles and carpeting. Made by Rtica, based in Stoney Creek, Ontario, the batts are installed just like fiberglass and make for an excellent fiberglass replacement choice.

But before ripping out that old fiberglass, it may be worth getting a professional to evaluate the integrity of your home's ductwork. With properly sealed ducts, any stray fiberglass slivers inside your walls shouldn't be able to get out. In the case of duct contamination, your best bet is to replace the entire system. Duct cleaning is also an option, but the EPA doesn't recommend it. If you do decide to opt for cleaning, the National Air Duct Cleaners Association offers a list of companies that can do the work.

CONTACTS: Fiberglass Information Network, www.sustainableenterprises.com/fin; Rtica, (905) 643-8669, www.rtica.com; National Air Duct Cleaners Association, (202) 737-2926, www.nadca.com.

What are the environmental benefits of an all-steel home? What other kinds of "green" homes are on the market today?

D. Hudson, Park City, UT

Homes on the cutting edge of building design today use steel framing, modular and panelized construction techniques, and energy-efficient insulation. The criteria for what is considered a "green" building varies, but most such structures offer good indoor air quality, reduced energy use, and resource conservation via the use of recycled, reused, or sustainably harvested virgin materials. Green buildings are often sited to minimize water use and runoff while taking full advantage of the sun for solar heating and/or shade for natural cooling. The initial costs of a green home might be more than for a traditional house, but the buyer's return on investment comes in the form of energy and maintenance cost savings over a lifetime.

While a wide range of construction materials passes the test as environmentally friendly, steel is king in the new generation of affordable green buildings. Besides its strength and resistance to weather and fire, steel is ultimately recyclable; two-thirds of all the steel in use in the United States today comes from recycled stock. Additionally, by framing houses with steel instead of wood, green builders save millions of trees every year.

Beyond steel, other materials such as adobe, straw bales, or "rammed earth" can make for some of the most energy-efficient and affordable structural elements. Each provides excellent insulation and can save on both materials and transportation costs if available and procured locally. Some designs include walls made by "stressed skin foam" panels, rigid foam that is sandwiched between oriented strand boards. More scaled-down green homes might use recycled newspaper for insulation in otherwise traditional walls. Also, salvaging materials such as old windowsills, floorboards, or light fixtures from existing or tear-down structures epitomizes the green motto "Reduce, Reuse, Recycle" while saving money.

Those inspired to build a green home on a limited budget today have a wealth of information at their disposal, notably a plethora of websites devoted to green building practices, techniques, and materials that offer free information online. Also, *Building Innovation for Homeownership*, a publication of the federally funded Partnership for Advancing Technology in Housing, profiles sixty-three award-winning, low-cost housing developments that incorporate materials and techniques on the cutting edge of green building. Meanwhile, the EPA offers free access to its online Energy Star database of green builders. The EPA site also includes a database of both lenders and utilities that offer special incentives to buyers and builders of energy-efficient homes.

CONTACTS: EPA Energy Star new homes partner locator, www.energystar.gov/index.cfm?fuseaction=new_homes_partners.showhomessearch; *Green Builder* magazine, www.greenbuilder.com; Green Building Resource Guide, www.greenguide.com; Center for ReSource Conservation, www.greenerbuilding.org; Partnership for Advancing Technology in Housing, www.pathnet.org.

How can I attract wildlife to my backyard?

Joshua Adam, Castine, ME

Go natural! You need an abundance of native plants as a food source, a water supply, and some form of shelter to encourage nesting. Birdbaths and fountains work well in the likely event that you don't have a natural stream or pond in your yard. And, while it may be tempting to remove dead or dying trees, woodpeckers depend on them, as do cavity-nesting birds such as owls and chickadees.

Rotting logs and mulch piles may seem to be eyesores, but they provide excellent habitat and nesting sites for small mammals, reptiles, and amphibians.

The more native perennials and annuals you plant, the more success you'll have attracting birds, butterflies, and other wildlife. And if your plants are truly native to your region, they will require little maintenance, as they have evolved to succeed there. They are adapted to local soil, rainfall, and temperature conditions and have developed natural defenses to many insects and diseases. Because of these traits, native plants will grow with minimal use of water, fertilizers, and pesticides. Wildlife species evolve with plants and use native plant communities as primary habitat, helping to preserve the balance of the local ecosystem.

Beyond the thrill of viewing wildlife out your window, there are other reasons to create a mini-refuge in your backyard. In the 1960s, Harvard biologist Edward O. Wilson likened backyards to wildlife "bridges" between protected areas that improved the chances of survival for many species. "The average American garden is home to hundreds of species of wildlife and acts as a vital corridor for migrating animals such as songbirds," agrees Jake Scott, an educator with the Backyard Wildlife Habitat Program of the National Wildlife Federation (NWF). Since 1973, NWF has been encouraging everyone from home owners to teachers to community leaders to plan their landscapes with the needs of wildlife in mind. Since the program started three decades ago, NWF has certified thousands of backyard habitats all over the United States and Canada as "wildlife friendly." Their program provides tools and resources that make getting started a snap.

The first step is to track down a good assortment of plants native to your region. NWF offers an easy-to-use online native plant guide that covers the continental United States and Alaska. Similarly, the Canadian Wildlife Federation offers *Backyard Habitat for Canada's Wild-*

life. If you're a novice, finding a store manager at a nearby nursery to serve as your personal advisor might be the best way to go, as there's nothing like local experience to make the most of attracting native wildlife to your yard.

CONTACTS: Canadian Wildlife Federation, (613) 599-9594, www.cwf-fcf.org; National Wildlife Federation Backyard Wildlife Habitat Program, (800) 822-9919, www.nwf.org/backyardwildlifehabitat.

What are some ways to maintain a "green" swimming pool?

Jim Humphey, North Andover, MA

The big health and environmental challenges with swimming pools are water and energy waste and overuse of chlorine. Chlorine is very irritating to the eyes and skin and can trigger breathing difficulties by also "stinging" the sensitive tissue of the lungs. The chemical's effects in a swimming pool are heightened when it comes into contact with sweat or urine. In fact, a recent Belgian study found a possible link between childhood asthma and exposure to chlorine by-products in indoor pools.

Zodiac Pool Care offers a system called Nature2 that doesn't do away with chlorine entirely but does greatly decrease the amount needed. It makes use of silver and copper to destroy bacteria and algae. Silver is a bactericide whose properties have long been known. Copper kills algae. When used together, they reduce chlorine needs by 90 percent. Another product, from ChlorFree, combines silver and copper with zinc, activated carbon, and other noninvasive materials to sanitize and control algae and bacteria and also greatly reduces the need for chlorine.

According to NSF International (formerly the National Sanitation Foundation), another substitute for chlorine is ozone, which is made from oxygen and does not degrade into harmful chlorinated by-products in a swimming pool. The Chlorine Free Products Association recently endorsed an ozone-only public pool built for the city of Fairhope, Alabama. The pool has been operating successfully since construction without the need for harmful additives. Ozone systems for residential pools are slowly becoming available. Sunshine Pool Products makes one that, according to owner Richard Barnes, should enable a completely chlorine-free environment if installed properly and at the right size for the dimensions of the pool.

Pool owners can save energy while still maintaining a pristine pool by using a timer to shut off the pump for at least twelve hours of the day. To hold in heat during the night, you should always use a pool cover, as almost all the heat loss occurs at the surface. By employing a bubble cover (sometimes called a solar cover), outdoor pools can also gain heat by absorbing 75 to 85 percent of the solar energy striking the pool's surface. A pool cover can also reduce water loss by 30 to 50 percent—and reducing water loss also reduces the amount of chemical water treatment required.

Besides that, the easiest way to save energy is to lower the thermostat on your pool's heater (if it has one) so that it heats the pool no higher than a minimally comfortable temperature. Every one-degree reduction in temperature can cut your energy use by between 5 and 10 percent.

CONTACTS: ChlorFree, (506) 665-0896, www.chlorfree.net; Chlorine-Free Products Association, (847) 658-6104, www.chlorinefreeproducts.org; NSF International, (800) NSF-MARK, www.nsf.org; Sunshine Pool Products, (801) 728-4520, www.sunshinepool.com; Zodiac Pool Care, (800) 937-7873, www.nature2.com.

What contaminants could be present in my well water, and how can I test for them?

Ruth Zandstra, Highland, IN

The EPA says that even when there is no human-caused pollution, groundwater, including well water, can contain a number of natural impurities that result from conditions in the watershed or in the ground. Water moving through underground rocks and soils can pick up magnesium, calcium, and chlorides. Some groundwater naturally contains elements such as arsenic, boron, selenium, or radon, a gas formed by the natural breakdown of radioactive uranium in soil. Whether these natural contaminants will cause health problems depends upon how much is present.

Well water can also be affected by improperly built or maintained septic systems nearby, leaking or abandoned underground storage tanks, storm-water drains that discharge chemicals into groundwater, chemical spills at local industrial sites, or improper disposal of pesticides, fertilizers, or animal manures. In 1999, nearly five hundred people were sickened and one child died in an outbreak of deadly *E. coli* bacteria at the Washington County Fair near Albany, New York. Health officials concluded that the water supply had been tainted when rainwater washed over cow feces from a cattle barn on the fairgrounds and ran into an underground aquifer tapped by the fair's wells.

About 15 percent of Americans obtain their drinking water from wells, cisterns, and springs. Unlike public water supplies, private wells in the United States are not regulated or regularly checked for contaminants. Therefore, home owners should periodically check their well water for the presence of potentially dangerous substances.

A good place to begin is with your local health department, which may provide free testing for contaminants or, at the very least, advice on how to proceed. If local testing is not an option, the EPA suggests that you find a state-certified lab through the yellow pages or online. Such labs can perform tests for bacteria, pesticides, nitrates, heavy metals, and other possible contaminants.

It is also possible to order specific tests from online labs, such as tdsmeter.com. The company can send a water testing kit with a prepaid envelope for mailing in samples. Results are then e-mailed back to the well owner. As another option, individuals can do their own well testing with home kits available from companies such as Promolife. Lab results can then be compared with public safety standards. If toxic levels are identified, you can discuss the results with your local health department.

CONTACTS: EPA, Safe Drinking Water Hotline, (800) 426-4791, www.epa.gov/safewater/pwells1.html; Promolife, (888) 742-3404, www.promolife.com; HM Digital, Inc., (800) 383-2777, www.tdsmeter.com.

What can be done to make office buildings more energy-efficient? So many leave thousands of lights on at night!

Deborah, Baltimore, MD

Office buildings are indeed the top energy guzzlers among commercial buildings in the United States, far above retail and service establishments and even manufacturing facilities.

The Department of Energy says that office building owners spend an average of $1.34 per square foot annually on electricity. Lights, office equipment, and heating/cooling systems account for about 90 percent of this expenditure. Lighting is clearly the main culprit, comprising 44 percent of all usage. Office equipment—computers, printers, copiers, fax machines, and telephone systems—accounts for about 23 percent.

Building managers can make a big difference by installing energy-efficient systems—from heating and cooling to lighting and waste disposal—but individual business owners and their employees can also have an impact by simply turning off lights and shutting down dormant machinery during nonworking hours.

According to Advanced Energy, a North Carolina–based nonprofit organization that monitors and analyzes energy use in commercial spaces, replacing older traditional fluorescent tubes with newer and more efficient compact fluorescent bulbs can save as much as 30 percent on electricity. And installing occupancy sensors so that lights go on and off as people enter and leave rooms can save an additional 5 percent. Building managers can save up to 15 percent on electricity bills by programming thermostats to trigger warming and cooling as needed during the workday while hibernating at night and on the weekends when buildings are mostly empty.

Periodically assessing and retooling heating and cooling systems can achieve additional energy savings. Any heating and cooling equipment older than a decade, for example, is probably ready for an upgrade to a newer more energy-efficient system. The federal government's Energy Star program, administered jointly by the EPA and the Department of Energy, rates the energy efficiency of lighting, office equipment, and heating/cooling systems from a wide range of manufacturers. Purchasing administrators can browse the Energy Star website to find out which models and systems will save a company the most money.

A handful of environmental groups are walking the talk via recent "green" retrofits of their office spaces. The National Audubon Society, Natural Resources Defense Council, and the Environmental Defense Fund, for example, have installed occupancy sensors and compact fluorescent lighting throughout their offices and in some cases have installed windows and configured their workspaces to make use of natural daylight instead of artificial light where possible. The Environmental Defense Fund's new San Francisco office is a state-of-the-art green building.

CONTACTS: Advanced Energy, (919) 857-9000, www.advancedenergy.org; Department of Energy's Energy Efficiency and Renewable Energy Program, www.eere.energy.gov; Energy Star, www.energystar.gov.

What are some ways to save paper at the office?

Wee Kheong, Singapore

The paperless office does appear to still be a distant dream. A recent University of California, Berkeley, study found that, worldwide, the amount of printed matter generated between 1999 and 2002 not only did not decrease—it grew by 36 percent. The average office worker prints and copies some ten thousand pages every year. The electronic age has not drastically cut paper consumption, but people can still do their part by replacing printed communications with virtual ones whenever possible.

Many companies use e-mail extensively now, for both interoffice exchanges and communication with customers. Attaching files to e-mails instead of printing out reports can eliminate reams of paper on a daily basis, as can posting information on company websites or intranets, private networks that use the same kinds of software as the public Internet but for internal use.

Beyond eliminating paper, offices have many options for reducing paper consumption. One very obvious strategy is to use both sides of every sheet, an approach that, if used religiously, can cut routine paper usage almost in half. Most office equipment can be set to default to double-sided printing. Also, workers should make use of the "print preview" function—which comes standard in most word processing and spreadsheet software—to prevent having to reprint pages due to errors discovered after the fact.

Sharing is another way to save paper. Notices or announcements can be posted in a few highly trafficked common areas instead of delivered to individual desks. Likewise, a single copy of a report can be circulated for editing to multiple employees. Meanwhile, more paper can be saved by printing only relevant pages instead of entire reports.

Copiers and printers that are in tune and running efficiently also help save paper. When copiers are not serviced regularly, they run out of toner and jam more often, causing more paper to be wasted. Printers suffer similar problems if ink cartridges wear out or paper trays are filled the wrong way.

Companies can help their employees save paper by instituting a formalized paper reduction campaign, including mandatory double-sided printing and copying, the scheduling of periodic equipment maintenance, and the reduction of paper-based forms. Outdated letterhead can be used as scratch paper or for internal memos. And office managers can make sure clearly marked paper recycling bins are available and can post "think before you print" and "think before you copy" signs in visible areas of the workplace.

CONTACTS: Department of Energy's Cutting Paper waste-minimization program, http://eetd.lbl.gov/paper/; INFORM's "Waste Reduction Tips for the Office," www.informinc.org/fact_office.php; California Integrated Waste Management Board, "Office Paper Reduction Quick Tips," www.ciwmb.ca.gov/bizwaste/officepaper/quicktip.htm.

5
PHANTOM LOADS AND ENERGY SUCKERS

Curing a Nation of Oil Addicts, One Turbine at a Time

In 1900, Americans had a choice between gasoline, electric, or steam cars; and their homes could be heated and lighted with coal, kerosene, alcohol, and even whale oil. There was a rich field of choice, but abundant, cheap oil arrived soon after that and (along with coal) has dominated our energy picture for more than a century. The good news is that we're at a similar crossroads, with alternative energy sources suddenly both technically and economically viable. Is it any wonder, as oil finally peaks, that we're getting seduced by wind turbines, vegetable-based biodiesel, and made-in-the-U.S.A. solar power? This may be a new American dream: drawing power from the sun, waves, plants, and even our waste. We might even learn to live with less, including a smaller carbon footprint.

Is the world running out of oil?

Allie Knopf, Kansas City, MO

Many experts say that evidence points to a declining world oil supply. According to petroleum geologist Colin Campbell, who has worked for Texaco, BP, Shell, and other major oil companies, world oil discovery peaked in the 1960s, while world production may have already peaked or will do so in a few years. Campbell predicts "the onset of a chronic long-term shortage" by 2010.

The Energy Information Administration (EIA) says the United States had 21.757 billion barrels of "proven" oil reserves as of January 2007, about 20 percent less than we had in 1990. "Proven" refers to estimated amounts that can be recovered in upcoming years with reasonable certainty. Outside the United States, nearly two-thirds of the world's proven oil reserves exist in the eleven countries that make up the Organization of Petroleum Exporting Countries (OPEC): Algeria, Indonesia, Iran, Iraq, Kuwait, Libya, Nigeria, Qatar, Saudi Arabia, the United Arab Emirates, and Venezuela.

The U.S. Geological Survey estimated in 2000 that 649 billion barrels of undiscovered oil and 612 billion barrels of oil reserve growth exist outside the United States. "Undiscovered" refers to oil located in places that haven't yet been drilled or explored; "oil reserve growth" refers to new discoveries near or in existing oil fields.

These estimates do not include oil sitting in storage facilities, such as the one billion barrels in the U.S. Strategic Petroleum Reserve, located underground in salt caverns along the Gulf of Mexico. It is the world's largest cache of emergency oil, with a provision of fifty-three days of import protection.

How much oil do we need anyway? The EIA's International Energy Outlook predicts that world demand will increase by 1.9 percent annually, up from 77 million barrels per day in 2001 to 121 million barrels per day in 2025, with much of the increase projected to occur in the United States, China, and other developing nations in Asia.

That's hardly a sustainable scenario given the likelihood of peak oil. Dr. Nancy Kete, director of the World Resources Institute's Climate, Energy and Pollution Program, says: "We must face the inescapable fact that the nation's environment, economy, national security, and oil resource base all point to the need for vast investments in energy efficiency and the rapid introduction of new, non-oil energy sources."

CONTACT: Energy Information Administration, (202) 586-8800, http://eia.doe.gov; U.S. Geological Survey, (888) ASK-USGS, www.usgs.gov; World Resources Institute, (202) 729-7600, www.wri.org.

Why is gasoline so much more expensive in Europe than in the United States?

Bo White, Chicago, IL

There are multiple components to gas prices: the cost of production and delivery, including the cost of crude oil to refineries and refinery processing costs; marketing and distribution costs; retail station costs; and taxes.

In one recent assessment, the price of crude oil accounted for about 43 percent of the cost of a gallon of regular grade gasoline; refining costs and profits comprised about 13 percent; distribution, marketing, and retail dealer costs and profits made up 13 percent; and federal, state, and local taxes accounted for approximately 31 percent of the cost.

Gasoline prices in countries such as the United Kingdom and Norway can sometimes reach five dollars per gallon–because of high taxes. According to the *Wall Street Journal*, taxes in the United Kingdom account for 80 percent of the pump price, while the Europe-wide average is between 60 and 70 percent.

In Germany, gasoline taxes account for a whopping 20 percent of all government revenues. Across Europe, such taxes have resulted in more fuel-efficient vehicles. According to John DeCicco of the Environmental Defense Fund, "The higher taxes have contributed to fuel efficiency that averages 30 percent higher [than U.S. levels]. However, they have not motivated ongoing conservation."

If gasoline taxes in the United States had the same effect on driving that cigarette taxes have had on some smokers, higher gas prices could provide the motivation for consumers to switch from, say, large SUVs to smaller, more fuel-efficient cars.

According to the *American Journal of Public Health*, every 10 percent increase in the price of cigarettes reduces smoking among pregnant women by 7 percent. The average American driver is certainly not as motivated to "do the right thing" as a mother-to-be, but it stands to reason that, like the effect of cigarette taxes, increased gas taxes might drive motorists to drive more fuel-efficient cars—and those tax revenues could be used to further promote fuel efficiency and develop alterative fuels.

CONTACT: Energy Information Administration, (202) 586.8800, www.eia.doe.gov; Environmental Defense Fund, (212) 505-2100, www.edf.org.

I've been hearing that wind power is going to play a significant role in our energy future. But doesn't it kill a lot of birds?

Dorothy Raffman, Norwalk, CT

Wind energy is zero-emissions energy, a renewable resource that many environmentalists and alternative energy proponents feel is one of our last, best hopes for staving off devastating climate change. According to the American Wind Energy Association (AWEA), the average wind turbine can prevent the emission of fifteen hundred tons of carbon dioxide each year.

Globally, wind energy has grown more than 500 percent since 1997, and 26 percent in 2006 alone. Total world capacity exceeded 74,000 megawatts by the end of 2006. The world wind leader is Germany, followed by the United States and Spain, but India and China are rapidly adding new wind turbines.

In the United States, there was huge growth in 2007: the wind energy industry installed 5,244 megawatts. The country's wind capacity grew by 45 percent in just one calendar year, with nine billion dollars in investment. "This is the third consecutive year of record-setting growth, establishing wind power as one of the largest sources of new electricity supply for the country," said AWEA executive director Randall Swisher.

The United States now has 16,818 megawatts of wind power in thirty-four states. U.S. wind farms generated an estimated 48 billion kilowatt hours of wind energy in 2008. A study predicts sixty-five billion dollars in U.S. wind investment through 2015.

In 2007, there were more than ninety thousand wind turbines operating worldwide. World wind leaders beyond Germany, the United States, and Spain include Austria, China, and India. A number of other countries, including the Netherlands, Italy, Japan, and Great Britain are right behind.

There are now wind energy installations in almost every American state west of the Mississippi and in many northeastern states too. Offshore wind has enormous growth potential as well. Germany, for instance, is building a 350-megawatt project (with seventy 5-megawatt turbines) anchored on the ocean floor off the island of Rügen. In Massachusetts, the Cape Wind project hopes to construct a $700 million, 420-megawatt, 130-windmill development that would stretch for five miles off Cape Cod, though it has drawn opposition from some residents, as has the German project, for fears that it will be an eyesore and could harm migrating birds.

But in truth, wind turbines get a bad rap for killing wildlife. High-profile examples such as at California's Altamont Pass—where outdated, oversized wind turbines kill some one thousand birds of prey each year—plague the growing wind power industry; even though more modern, better-sited wind farms kill far fewer birds.

According to researcher Wallace Erickson, birds face daily threats far more lethal than wind turbines. For instance, he reported that between five hundred million and one billion birds are killed annually in the United States alone from collisions with man-made structures—including communications towers, buildings, and windows—and contact with power lines. Hunting, cat predation, pesticides, commercial fishing operations, oils spills, and cars and trucks also take a heavy toll. Wind power advocates say the relatively small impact of windmills on bird populations has to be put in perspective. Wind turbines caused less that 1 percent of the total number of human-caused bird deaths in Erickson's study.

But even a few bird deaths are too many if they can be avoided. The American Bird Conservancy advises that lighting on turbines should be minimized, tension wires and lattice supports should be avoided, and wind turbine power lines should be placed underground whenever possible. Modern wind towers are being designed to prevent birds from perching on them. Also, the turbine blades rotate much more slowly than they did on earlier designs.

CONTACTS: American Bird Conservancy, www.abcbirds.org; American Wind Energy Association, www.awea.org; Cape Wind project, (617) 904-3100, www.capewind.org; European Wind Energy Association, (011) 32-2-546-1940, www.ewea.org; National Audubon Society, www.audubon.org.

I'm "pro-solar" all the way for the sake of the environment, but solar power has not historically been very cost-effective. What coming innovations will bring costs down to make solar competitive with other energy sources?

Will Proctor, Richmond, VA

The prospect of pollution-free power from the sun is appealing, but to date the low price of oil combined with the high costs of developing new technology have slowed the growth of solar power in the United States and beyond. At twenty-five to fifty cents per kilowatt-hour, solar power costs as much as five times more than conventional fossil-fuel-based electricity, though rising utility rates are evening things up. The dwindling supply of polysilicon, the element found in traditional photovoltaic cells, is not helping. Prices per watt rose in 2006.

Gary Gerber of Sun Light and Power in Berkeley, California, points out that Ronald Reagan moved into the White House in 1980 and removed the solar collectors from the roof that Jimmy Carter had installed. Soon after, tax credits for solar development disappeared and the industry plunged "over a cliff."

Federal spending on solar energy picked up under the Clinton administration but trailed off again once George W. Bush took office. Growing climate change worries and high oil prices have forced the Bush administration to reconsider its stance on alternatives like solar, and proposed funding rose 80 percent in 2007 to $148 million.

Despite obstacles, solar is making a lot of progress. Worldwide, energy from the sun is "shining bright," according to the Worldwatch Institute. Since 2000, global production has increased sixfold. It's now the world's fastest-growing energy source.

The leading solar manufacturer is Japan, with China growing rapidly. The biggest customer is Germany, which added 1,100 megawatts in 2006 (half of all the solar worldwide).

Solar engineers are working hard to get costs down and expect it to be competitive with fossil fuels within twenty years. One technological innovator is California-based Nanosolar, which replaces the silicon used to absorb sunlight and convert it into electricity with a thin film of copper, indium, gallium, and selenium (CIGS). Says Nanosolar's Martin Roscheisen, CIGS-based cells are flexible and more durable, making them easier to install in a wide range of applications. Roscheisen expects he will be able to build a 400-megawatt electricity plant for about a tenth of the price of a comparable silicon-based plant. Other companies making waves with CIGS-based solar cells include DayStar Technologies and Miasolé.

Another recent innovation is the so-called spray-on cell, as made by Massachusetts-based Konarka. The composite can be sprayed like paint onto other materials, where it can harness the sun's infrared rays to power cell phones and other portable or wireless devices. Some analysts think spray-on cells could become five times more efficient than the current photovoltaic standard.

Environmentalists and mechanical engineers aren't the only ones bullish on solar these days. According to the Cleantech Network, venture capitalists poured some $100 million into solar start-ups of all sizes in 2006 alone, and the business is growing steadily. Given the venture capital community's interest in relatively short-term returns, it's a good bet that some of today's promising solar start-ups will be tomorrow's energy behemoths.

CONTACTS: DayStar Technologies, www.daystartech.com; Konarka, www.konarka.com; Miasolé, www.miasole.com; Nanosolar, www.nanosolar.com; PowerFilm, www.powerfilmsolar.com; Sun Light and Power, www.sunlightandpower.com.

What's happening with wave power?

Tina Cook, Naples, FL

As any board or body surfer will tell you, the ocean's tidal currents pack considerable wallop. So why wouldn't it make sense to harness all that formidable power, which is not unlike that of the rivers that drive hydropower dams or the wind that drives wind turbines?

The concept is simple, says John Lienhard, a University of Houston mechanical engineering professor: "Every day the moon's gravitational pull lifts countless tons of water up into, say, the East River or the Bay of Fundy. When that water flows back out to sea, its energy dissipates and, if we don't use it, it's simply spent." According to Energy Quest, an educational website of the California Energy Commission, the sea can be harnessed for energy in three basic ways: using wave power, using tidal power, and using ocean water temperature variations in a process called "ocean thermal energy conversion."

In harnessing wave power, the back-and-forth or up-and-down movement of waves can be harnessed, for example, to force air in and out of a chamber to drive a piston or spin a turbine that can then power a generator. Some systems in operation now power small lighthouses and warning buoys. Tidal energy, on the other hand, involves trapping water at high tide and then harnessing its energy as it rushes out and drops in its change to low tide. This is similar to the way water makes hydroelectric dams work. Already some large installations in Canada and France generate enough electricity to power thousands of homes.

One tidal system uses the temperature differences between deep and surface water to extract energy. An experimental station in Hawaii hopes to develop the technology and someday produce large amounts of electricity on par with the cost of conventional power technologies.

Partisans say that ocean energy is preferable to wind because tides are constant and predictable, and that water's natural density requires fewer turbines than are needed to produce the same amount of wind power. Given the difficulty and cost of building tidal arrays at sea and getting the energy back to land, however, ocean technologies are still young and mostly experimental. But as the industry matures, costs will drop and some analysts think the ocean could power nearly 2 percent of U.S. energy needs.

Several companies now work at the cutting edge of ocean power technology. Pelamis Wave Power in Scotland has a wave system called Pelamis that it hopes to install in waters off California's wave-battered Central Coast. And Aqua Energy in Seattle has installations off the coasts of Oregon, Washington, and British Columbia and is in talks with utilities about providing the Pacific Northwest with hundreds of megawatts of ocean energy within the next decade.

Tidal energy pioneers are also hard at work on the Atlantic coast. The New Hampshire Tidal Energy Company is developing tidal power in the Piscataqua River between New Hampshire and Maine. And a company called Verdant Power is providing Long Island City, New York, with electricity through tidal river turbines and has begun installation of tidal power systems in New York City's East River.

CONTACTS: Aqua Energy (Finavera Renewables), www.finavera.com/wave; Pelamis Wave Power, www.pelamiswave.com; Verdant Power, www.verdantpower.com.

What are the pros and cons of switching from oil to plant-based biofuels?

Jim Dand, Somerville, MA

Since biofuels are derived from agricultural crops, they are inherently renewable. And our own farmers produce them domestically, reducing our dependence on unstable foreign sources of oil. Additionally, ethanol and biodiesel emit less particulate pollution than traditional petroleum-based gasoline and diesel fuels. Looked at in one way, they also do not contribute to global warming, since they only emit back to the environment the CO_2 that their source plants absorbed out of the atmosphere in the first place.

The transition from fossil fuels to biofuels would be relatively simple compared to other forms of renewable energy (like hydrogen, solar, or wind). If you have a diesel or flex-fuel vehicle, it's already biofuel compliant, and concentrations of biodiesel can be burned in home heating oil tanks. Worldwide production of biofuels rose 28 percent in 2006.

Despite these upsides, however, biofuels are not a simple cure for our addiction to petroleum. A wholesale societal shift from gasoline to biofuels, given the number of gas-only cars already on the road and the lack of ethanol or biodiesel pumps at existing filling stations, would take some time.

Another major hurdle for widespread adoption of biofuels is the challenge of growing enough crops to meet demand, something skeptics say might well require converting just about all of the world's remaining forests and open spaces to agricultural land. "Replacing only

5 percent of the nation's diesel consumption with biodiesel would require diverting approximately 60 percent of today's soy crops to biodiesel production," says Matthew Brown, an energy consultant and former energy program director at the National Conference of State Legislatures. "That's bad news for tofu lovers."

Ethanol is primarily in use today as an octane-boosting fuel additive, but it can also be used in 85 percent concentrations in "flex-fuel" versions of such vehicles as the Ford Explorer and Chevy Silverado. In order to stimulate production, the United States offers generous tax-based subsidies to farmers who grow crops for ethanol.

But according to UC Berkeley researcher Tad Patzek, "People tend to think of ethanol and see an endless cycle: Corn is used to produce ethanol, ethanol is burned and gives off carbon dioxide, and corn uses the carbon dioxide as it grows. But that isn't the case. Fossil fuel actually drives the whole cycle."

Another dark cloud is energy yields. Cornell University researcher David Pimentel, who has collaborated with Patzek, concludes that producing ethanol from corn requires 29 percent more energy than the end product itself is capable of generating. He found similarly troubling numbers in making biodiesel from soybeans. "There is just no energy benefit to using plant biomass for liquid fuel," says Pimentel.

We're likely to see a combination of sources—from wind and ocean currents to hydrogen, solar, and, yes, some use of biofuels—meeting our future energy needs. And probably the largest single "alternative fuel" available to us is energy conservation.

CONTACTS: Earth 911, "Energy Conservation Factsheet," www.earth911.org/energy/energy-costs-and-conservation-facts; Ecology Center, "Biodiesel FAQ," www.ecologycenter.org/factsheets/biodiesel.html.

I'm concerned about all the talk of using hydrogen for fuel. Isn't hydrogen what caused the *Hindenburg* to explode back in the 1930s?

Doug, via e-mail

The explosion of the blimp *Hindenburg* in Lakehurst, New Jersey, in 1937 killed thirty-six people and was one of the worst air disasters of the period, but hydrogen was arguably not the culprit. Addison Bain, a National Aeronautics and Space Administration (NASA) researcher, investigated the *Hindenburg* crash in 1997. He concluded that, while the *Hindenburg* did use hydrogen for buoyancy, the cause of the accident was an electrostatic charge that ignited the blimp's highly flammable waterproof skin, made from a mixture of lacquer and metal-based paints that Bain likened to rocket fuel. Bain's theory has been disputed, however.

Others argue that a spark ignited hydrogen that was leaking from the ship. But witnesses described the fire as very colorful, and hydrogen burns without much of a visible flame. Whether hydrogen caused or simply contributed to the ensuing blaze, it's certainly dangerous and highly flammable.

But gasoline is much more flammable than hydrogen, which is also fourteen times lighter than air. When it catches fire it disperses and extinguishes quickly. Gasoline, on the other hand, is heavier than air and stays flammable much longer.

Besides being less flammable than gasoline, hydrogen has many other benefits. It is nontoxic, an advantage it has over any petroleum-based fuel. Further, the chemical reaction that produces electricity from hydrogen in fuel cells produces no harmful pollutants and emits only pure, potable water as well as heat that can be recaptured for other uses. The combustion of gasoline and other automotive fuels leads to acid rain, smog, and global warming, among other environmental problems.

Despite its benefits, the widespread adoption of hydrogen as an automotive fuel is not yet close at hand. Techniques for producing, storing, and transporting hydrogen have to be standardized and costs reduced substantially. Some hydrogen proponents see a future where hydrogen will fuel vehicles at service stations, as is now done with gasoline; others see a future in which people fuel their cars at home from appliances that make hydrogen from electricity or, further along, from solar energy.

In 2003, the Bush administration committed $1.2 billion to a hydrogen initiative, though the figure was somewhat misleading (it included funds in existing programs). The White House proclaimed, "The first car driven by a child born today could be powered by fuel cells." But many more billions of dollars, mostly from the auto industry, which is already fielding sophisticated fuel-cell cars in test programs, will have to be invested before that dream becomes reality.

CONTACTS: Energy Independence Now, www.energyindependence now.org; White House, "Hydrogen Economy Fact Sheet," www.white house.gov/news/releases/2003/06/20030625-6.html.

What exactly are fuel cells, and what can they power that will end or reduce our dependence on oil and gasoline?

Alex Tibbetts, Seattle, WA

First developed as a power source for NASA's Apollo missions, fuel cells convert hydrogen and oxygen into usable electricity, with heat and water as by-products. Gasoline engines like those found under the hoods of today's cars harness energy by burning fossil fuels; fuel cells derive power much more efficiently via chemical reactions between hydrogen and oxygen.

Fuel-cell technology can be used to run everything from laptop computers to power plants. Cities in the United States, Europe, and China currently operate public bus fleets powered by hydrogen fuel-cell engines. King County in Washington State is using fuel cells to power its new water-treatment plant. And eight of the world's top automakers are developing prototype cars and trucks powered by fuel cells. The current leaders are General Motors, which offers both the cutting-edge Sequel and small fleets of SUV-style Equinox vehicles, and Honda, which is beginning to lease FCX Clarity cars to a few select customers.

Governments and automakers are supporting the research and development with various investments, grants, and subsidies. In 2002, President Bush launched the FreedomCAR program, a public-private partnership between the Department of Energy and the "Big Three" automakers, to fund development of fuel-cell technologies for American cars and trucks. A year later, the White House announced the creation of the Hydrogen Fuel Initiative to offer support for a hydrogen-refueling network throughout the United States and beyond.

But environmental critics are suspicious of the Bush administration's motives, especially since the Energy Department's priorities lie with generating hydrogen from coal or nuclear power rather than from sustainable sources like solar or wind power. But in a positive sign, the United States and European Union have agreed to work jointly on fuel-cell development initiatives.

The promise of a transportation sector powered by hydrogen fuel cells is appealing for economic and political reasons as well as for environmental ones. Besides the well-understood negative impacts of fossil-fuel emissions on our air, water, and health, experts are predicting that the peak of oil production will soon be reached, with remaining supplies largely in the volatile Middle East.

Despite their promise, though, fuel cells are not about to take over anytime soon. "Fuel-cell vehicles will not make a significant national impact for at least two decades," says Jason Mark of the Energy Foundation. But Mark remains bullish on the future of fuel cells. "Given the pressing economic and environmental risks posed by automobile travel, we can't afford to pass up their tremendous long-term potential."

CONTACTS: FreedomCAR program, (877) 337-3463, www.eere.energy.gov/vehiclesandfuels; Hydrogen Fuel Initiative, www.whitehouse.gov/news/releases/2003/02/20030206-2.html; Plug Power, (518) 782-7700, www.plugpower.com; UTC Fuel Cells, (860) 727-2200, www.utcfuelcells.com.

What do you think of those "waste-to-energy" plants used by cities to generate power?

Christine Ramadhin, Queens, NY

Waste-to-energy (WtE) facilities, which generate power by burning trash, have been in widespread operation in the United States and Europe since the 1970s and are something of a mixed blessing. They process garbage without adding to already stressed landfills and with the added benefit of contributing electricity to the power grid. But they also generate toxic pollution, usually as a result of burning vinyl and plastics.

WtE facilities evolved out of basic incinerator technology that simply burns trash and reduces it to ash and smoke. Waste-to-energy plants instead use the garbage to fire a huge boiler. When the garbage "fuel" is burned, it releases heat that turns water into steam. The high-pressure steam turns the blades of a turbine generator to produce electricity.

In the United States and Europe, environmental laws regulate WtE plants, typically requiring them to use antipollution devices to keep both harmful gases and particulate pollution (fine bits of dust, soot, and other solid materials) out of the air. However, the particles captured are then mixed with the ash that is removed from the bottom of the plant's furnace when it is cleaned. Environmentalists contend that this toxic ash, which can include dangerous heavy metals, may actually present more of an environmental problem than the airborne emissions themselves, as it usually ends up in landfills where it can leak into and contaminate soil and groundwater.

According to Greenpeace International, WtE facilities are also among the largest sources of dioxin emissions in industrialized countries. Dioxin is a by-product of burning polyvinyl chloride (PVC) and other plastics and has been linked to cancer and other health problems. Greenpeace advocates phasing out WtE facilities in favor of improving recycling rates that reduce the waste stream in the first place.

Currently about six hundred WtE facilities are in operation around the world. According to the National Solid Wastes Management Association, the United States is home to ninety-eight such plants operating in twenty-nine states. These facilities manage about 13 percent of America's total trash output. In Canada, where landfill space is abundant, WtE has failed to catch on, and there are only a few such facilities across the country. WtE is more popular in smaller, technologically advanced countries such as Japan, Sweden, Denmark, France, and Switzerland, where landfill space is at a premium.

Recent improvements in the energy efficiency and environmental impact of WtE facilities means that the technology will probably continue to play a role for years to come, especially as developing countries start to use it.

CONTACTS: Greenpeace incineration campaign, www.greenpeace.org/international/campaigns/toxics/incineration; National Solid Wastes Management Association, www.nswma.org.

Despite all the talk about wind and solar power, isn't hydro the largest renewable source?

Bianca Hoffman, Bridgeport, CT

Hydropower—energy generated from water flowing through turbines in dams—is definitely still the king of renewables. Globally, hydropower generates 20 percent of the world's electricity. In Canada, which is the world's largest generator of hydropower, over 60 percent of electricity comes from the power of water. Norway gets almost 99 percent of its electricity from hydropower, and New Zealand is close behind at 75 percent. In the United States, about 10 percent of all electricity—enough to power thirty-five million homes every year—comes from hydropower.

Hydro has been part of America's energy mix since the 1880s, when the world's first hydroelectric plant began operation on the Fox River in Appleton, Wisconsin. By the 1940s, hydropower accounted for about 40 percent of America's energy needs.

Many environmentalists still cheer hydropower as the only major source of electricity that is renewable and nonpolluting. Unlike energy generated from fossil fuels, hydropower plants do not emit the waste heat and gases that are major contributors to air pollution, global warming, and acid rain. Nor do they require the environmentally destructive mining and drilling needed to acquire coal, natural gas, and oil.

But although hydropower does not generate pollution, per se, it has hurt salmon populations on both U.S. coasts. That's why a move to remove hydropower dams is slowly gaining ground. One of the first such projects was in Maine, where a major utility agreed to remove three dams on the Penobscot River and its tributaries in order to restore declining populations of wild Atlantic salmon. Environmentalists are calling for similar measures in the Pacific Northwest to save dwindling populations of coho and chinook salmon.

In China and India, large controversial dam projects have flooded huge areas of land and forced the relocation of whole communities of people.

Hydro is one of many renewables with both unrealized potential and significant challenges. Despite the promise of renewables, though, the United States still generates more than 90 percent of its energy from nonrenewable and polluting sources like coal and petroleum—and there is talk of a nuclear "revival" despite the potential dire consequences of a nuclear accident or terrorist act. Finding more efficient ways to harness solar energy is a top priority for many environmentalists, especially since the earth receives more energy from the sun in just one hour than the world uses in a whole year.

CONTACTS: Department of Energy, Renewable Energy program, www.eia.doe.gov/fuelrenewable.html; National Hydropower Association, www.hydro.org.

Nuclear power seems like such a clean and cost-effective alternative to burning fossil fuels. Why are so many environmentalists against it?

Paul Franklin, Missoula, MT

Nuclear power doesn't release significant amounts of the CO_2 that causes global warming or the airborne pollutants that cause respiratory harm. But the technology does have a serious downside: It generates radiation that can cause a host of genetic abnormalities, notably cancer. The lymphatic system, bone marrow, intestinal tract, thyroid, and the female breast are most vulnerable to the effects of radiation, especially in children and adolescents.

When most people think of the dangers of nuclear energy, they think of the highly publicized accidents that occurred at Three Mile Island in Pennsylvania in 1979 and Chernobyl in Ukraine in 1986. Studies have turned up very little measurable environmental damage from the Three Mile Island accident, which was quite minor compared to Chernobyl. Although casualty figures are in dispute, Ukrainian officials blame that accident for at least forty-three hundred deaths beyond the thirty that occurred during the meltdown and immediately afterward. And there is little disagreement that the accident caused as much as a hundredfold increase in thyroid cancers among children in Ukraine and in nearby Russia and Belarus. The Ukrainian Ministry of Health says that 2.32 million people there, including 452,000 children, have been treated for radiation-linked illnesses, including thyroid cancers and blood cancers like leukemia.

A nuclear accident, however, does not have to occur for radiation to escape and pose a health threat. The waste from power plant operations is also radioactive, and already the U.S. nuclear industry has left behind a legacy of nearly one hundred thousand tons of it. Scientists have not yet found a way to store nuclear waste—which stays radioactive for thousands of years—such that they can guarantee it won't harm people, even when it is buried miles below ground. The nuclear industry and the Bush administration are proposing that Yucca Mountain in Nevada be a central repository for the nation's nuclear waste, but transporting the waste there from widely dispersed locations poses significant risks in itself.

Some environmentalists are advocating the use of more nuclear energy, but primarily as a stopgap measure to stave off global warming. Among them is scientist and author James Lovelock, an architect and an intellectual leader in the modern environmental movement. Lovelock argues that global warming is happening too fast and that renewable sources of energy like wind and solar are not developing fast enough to reverse that trend. "I don't see nuclear as the ultimate solution," he said. "I see it as a kind of medicine, which is an unpleasant medicine in some ways that we have to take while we're curing ourselves of fossil fuels."

In 2004, in the pages of the British newspaper the *Independent*, Lovelock entreated his "friends in the movement to drop their wrongheaded objection to nuclear energy." That drew a response from Stephen Tindale, executive director of the British Greenpeace group. "Nuclear creates enormous problems," Tindale said, including "waste we don't know what to do with; radioactive emissions; unavoidable risk of accident; and terrorist attack."

CONTACTS: Environmentalists for Nuclear, www.ecolo.org; International Chernobyl Research and Information Network, www.chernobyl.info; Nuclear Information and Resource Service, www.nirs.org.

What's wrong with Nevada's Yucca Mountain as a safe place to store nuclear waste?

Vinka Lasic, Cleveland, OH

Since the 1980s, the Department of Energy has been pushing to open Nevada's Yucca Mountain as a nuclear waste storage facility. In 2002,

George W. Bush signed into law a plan to make the site the central repository for the spent nuclear fuel and high-level radioactive waste presently stored in separate locations in forty-three states. Yucca Mountain is ninety miles northwest of Las Vegas, and many environmentalists, area residents, and local and state officials believe it is dangerously unsuitable for nuclear waste storage.

Judy Treichel, executive director of the Nevada Nuclear Waste Task Force, cites "numerous reasons" to abandon the site, including the fact that both independent and state-sponsored scientists have determined that it's geologically active and located near active volcanoes. The Las Vegas- and Reno-based organization Citizen Alert says the proposed site lies on thirty-two known fault lines and has a history of rising groundwater. If the facility were to get flooded, the groundwater could be contaminated with hazardous materials.

John Hadder, Citizen Alert's northern Nevada coordinator, is concerned about the danger of transporting the nation's nuclear waste to Yucca Mountain from the many distant locations where it now sits. The waste would arrive by truck, and six to seven shipments of the hazardous material would be made daily for the next thirty years. Such a transportation system has inherent dangers, such as spills due to accidents and the possibility of terrorist attacks, according to the National Safety Council. Citizen Alert also worries that the communities through which the vehicles pass would suffer economically if the plan goes through.

Most Nevadans, including those in Native American communities, are dead set against their state becoming the nation's nuclear waste repository. When George W. Bush became president in 2000, he said he would base his decision on whether or not to allow nuclear waste storage at the site based on "sound science." Two years later, despite recommendations to the contrary from federal scientists and the Government Accountability Office, and after heavy lobbying by the nuclear power industry, he approved the plan, much to the dismay of Nevada's congressional delegation.

Currently, a handful of lawsuits challenging the plan are underway, and Nevadans are scrambling to propose alternative scenarios for handling nuclear waste. Meanwhile, Yucca Mountain could start accepting nuclear waste from across the country as soon as 2010.

CONTACTS: Nevada Nuclear Waste Task Force, www.nvantinuclear.org; EPA's Yucca Mountain information, www.epa.gov/radiation/yucca; National Safety Council, (630) 285-1121, www.nsc.org; Nuclear Information and Resource Service, (301) 270-NIRS, www.nirs.org/radwaste/yucca/yuccahome.htm.

What is geothermal heating and cooling, and how is it environmentally friendly?

John Moran, Cranston, RI

Geothermal (sometimes called "geoexchange") heating and cooling is a technology that relies primarily on the earth's natural thermal energy, a renewable resource, to heat or cool a house.

In winter or in colder climates, the earth's natural heat is collected through a series of pipes, called a loop, installed underground or sometimes in a pond or lake. Water circulating in the loop carries the heat to the home, where an indoor system using compressors and heat exchangers concentrates the earth's energy and releases it at a higher temperature. In a typical system, duct fans distribute the heat to various rooms. These systems can also provide all or part of a household's hot water, according to the Geothermal Heat Pump Consortium, a trade organization.

In summer or in warmer climates, the process is reversed in order to cool the home. Excess heat is drawn out, expelled to the loop, and absorbed by the earth. Thus the system provides cooling in much the same way that a refrigerator keeps its contents cool—by drawing heat from the interior rather than injecting cold air from the exterior. The only additional energy these systems need, other than the heat from the earth's surface, is a small amount of electricity to power the pumps that circulate the collected heating or cooling throughout the home.

"It's a truly renewable system requiring a minimal amount of energy," says Lisa McArthur of the International Ground Source Heat Pump Association, another trade group. "The temperature underground is constant year round (low forties in the northern U.S. to the low seventies in the South). If a home needs to be heated in the winter or cooled in the summer, the energy source is in one's own backyard," she says.

Depending on the size and quantity of pumps needed, home owners can expect to pay a few thousand dollars more for installation than for a conventional fossil-fuel system. But with geothermal, home owners enjoy reduced energy bills, high reliability, and long life. "There is always initial sticker shock, but our clientele is more concerned with the environment and long-term use rather than the initial bottom line," says Scott Jones of ECONAR, a Minnesota-based heat pump producer.

The Department of Energy says that geothermal technology can reduce energy costs up to 60 percent compared to traditional furnaces, yielding a two- to ten-year payback. Subsidies and tax incentives, which vary from state to state, can make the systems even more affordable. Home owners can check with the free online Database of State Incentives for Renewable Energy to see if their state provides any such incentives.

CONTACTS: Database of State Incentives for Renewable Energy, www.dsireusa.org; ECONAR, (763)241-3110, www.econar.com; Geothermal Heat Pump Consortium, (202) 558-7175, www.geoexchange.org; International Ground Source Heat Pump Association, www.igshpa.okstate.edu.

I'd like to start saving more energy in my home. Do you have any tips?

Mitch Rochelle, Carson City, NV

A University of Michigan study estimates that the average American household could reduce its energy bills by 65 percent and, over the home's lifetime, save $52,000 if it maximized energy efficiency.

One place to start is household appliances. Washers and dryers generate lots of heat, so in a warm climate they should be in sealed-off rooms so as not to exacerbate air-conditioning demand. Dishwashers and ovens should, if possible, be run in the early morning or evening to minimize heat buildup. On older refrigerators, vacuum the coils at the back of the unit regularly to keep them clean and free from dirt and dust, since that will compromise efficiency.

Repairing old appliances can improve energy efficiency somewhat, but replacing them with new models that comply with the federal Energy Star standards can reduce household energy costs by 20 percent. Consumers should remember that getting the right size unit installed professionally is essential to getting the most from new appliances.

Air-conditioning and heating need not take such a huge bite out of America's energy dollar. According to the Natural Resources Defense Council, if your air conditioner is more than eight years old, it's a good candidate for replacement. If your furnace or boiler is old or simply inefficient, the best solution is to replace it with a modern high-efficiency model. And to keep heating bills to a minimum, install a programmable thermostat, and schedule it to trigger heat only during the hours that you're home.

Many older homes are poorly sealed and lack insulation, sending energy bills skyrocketing. Also, it is common to find gaps between duct joints, whether a home is new or old. Seal and insulate ducts that are exposed in areas such as your attic or crawl space to improve your system's efficiency. According to the EPA, by properly sealing air leaks and adding insulation, you can improve comfort and cut your energy bills by up to 10 percent.

For a do-it-yourself assessment of your home's potential energy efficiency, check out the Department of Energy's Home Energy Saver website. Special software enables users to input information about their homes and then learn how much energy (and money) could be saved by insulating the attic or installing double-glazed windows. Indeed, with rising fuel costs, there's no time like the present to save energy in your home.

CONTACTS: Energy Star program, (888) STAR-YES, www.energystar.gov; Home Energy Saver website, http://hes.lbl.gov; Natural Resources Defense Council, (212) 727-2700, www.nrdc.org.

I've heard that tankless water heaters are more energy efficient than traditional water heaters. How do they work?

Felipe Gomez, Flagstaff, AZ

In a conventional water heater, thirty to sixty gallons of water sit in the tank, constantly being heated and reheated, even when no hot water is in use. The heat from the tank keeps dissipating into the air, creating "standby heat loss." This constant energy waste adds up, and can constitute 10 to 20 percent of a household's heating costs.

Unlike traditional water heaters, tankless water heaters (also known as "demand" or "instantaneous" water heaters) heat the water only as it is used, thus eliminating standby heat loss and minimizing energy usage. Cold water travels through a pipe to the unit, where it passes over a gas or electric heating element in a thin enclosure. This exposes a lot of the water's surface to the heating element, thus enabling it to heat up quickly. The element only operates when the hot-water faucet is turned on. These heaters are also small and thus space saving, and can be attached to a wall or put under the sink or in a closet.

First put into widespread use in Japan and Europe, tankless water heaters began appearing in the United States about twenty-five years ago. While they do cost more than double the price of conventional water heaters (top-of-the line, high-capacity residential tankless models sell for up to a thousand dollars), a typical tankless unit lasts more than twenty years, compared to the ten-year lifespan of a conventional water heater. Also, consumers can quickly make up the difference through energy savings.

While a constant supply of hot water is available through a tankless system, the flow rate may be somewhat limited, depending upon the needs of your household. Typically, a tankless water heater provides a flow of two to four gallons per minute. As with many tank heaters, simultaneous use of hot-water appliances can affect the flow rate. Water-hungry appliances like dishwashers and washing machines may need to be operated at separate times. Alternatively, a second water heater can be installed at a high-demand location. Gas-fired heaters tend to have higher flow rates and are less expensive than electric models. Leading tankless water heater manufacturers include Bosch, PowerStar, and Ariston, and the units are available at most big appliance and home superstores as well as through Controlled Energy Corporation, Tankless Water Heaters Direct, and several other dealers.

CONTACT: Controlled Energy Corporation, (800) 503-5028, www.boschhotwater.com; Energy Efficiency and Renewable Energy Office, (800) DOE-3732, www.eere.energy.gov/consumer/your_home/water_heating/index.cfm/mytopic=12760; Tankless Water Heaters Direct, (802) 583-5510, www.tanklesswaterheatersdirect.com.

What is biomass energy, and where in the world is it used?

Kourosh Khazaii, Vancouver, BC

Biomass energy is power generated by burning any organic plant matter, including wood, and that makes it perhaps the earliest source of fuel. Wood is by far the most widely used biomass energy source, but other plants are also used, as are residues from agriculture or forestry and the organic components of municipal and industrial wastes.

Environmentalists like biomass energy because it is fundamentally renewable and has the potential, if widely used, for greatly reducing the greenhouse gas emissions that contribute to global warming. While the burning of biomass fuels generates CO_2, new plants grown for biomass remove CO_2 from the atmosphere. So as long as biomass energy sources continue to be replenished, their net emissions will be zero.

Biomass, because it is available on a recurring basis, is the world's most plentiful fuel source, and it is second only to hydropower in efficiency. Farmers around the world are now cultivating fast-growing trees and grasses specifically for biomass energy use.

Developing countries in Asia, Latin America, and Africa are currently the primary users of biomass, depending on it for almost a third of total energy use. By contrast, the United States uses biomass for only 4 percent of its total energy supply.

Australia is generally recognized as the world leader in developing biomass projects, due to the close cooperation there between government agencies, research facilities, and industry. Britain is also working on some significant biomass projects, including the establishment of power stations fueled by fast-growing crops.

The International Energy Agency reports that biomass has the potential to supply 40 percent of the world's energy needs. Studies by the Shell International Petroleum Company and the UN-based Intergovernmental Panel on Climate Change are equally if not more optimistic and project that biomass could satisfy between one-quarter and one-half of the world's demand for energy by the middle of this century. This projection implies a world full of "biorefineries," where plants provide many of the materials we now obtain from coal, oil, and natural gas.

Looking ahead, some analysts have begun to talk about a "carbohydrate economy" in which plants would be a major source of not only electricity and fuel but also construction materials, clothes, inks, paints—even industrial chemicals.

CONTACTS: Biomass Energy Research Association, (800) 247-1755, www.beraonline.org; Intergovernmental Panel on Climate Change, www.ipcc.ch; International Energy Agency, (011) 33-1-40-57-65-00, www.iea.org; Shell International Petroleum Company, (888) GO-SHELL, www.shell.com.

What incentives are in place for home owners and businesses to install renewable energy systems?

Kelly Nemi, Sacramento, CA

Several state and municipal governments are trying to stimulate demand for alternative energy by offering cash incentives to companies and home owners that install solar electric (photovoltaic) systems, fuel cells, small wind turbines, solar thermal systems for heat and hot water, and other renewable energy technologies.

Anaheim, California's public utility, for instance, is encouraging residential and business customers to install photovoltaic systems by offering rebates of $4 per watt up to $7,000 total for residential systems and $50,000 for industrial installations. The state of Indiana's Alternative Power and Energy Grant program will help businesses, nonprofit organizations, and units of local government (such as schools) with the costs of installing solar, wind, fuel cell, geothermal, hydropower, alcohol fuel, waste-to-energy, and biomass energy technologies. The state will pay up to 30 percent of the project cost, or $30,000, whichever is less. And New Jersey's Clean Energy Rebate Program offers between 30 cents and $5.50 in rebates per watt for commercial or residential solar electric systems, depending upon size. Connecticut has similar programs through its Clean Energy Fund.

These are just a few examples. The Database of State Incentives for Renewable Energy (DSIRE) is a good place to find incentives that work for you. The website features a clickable U.S. map for consumers to access detailed information on what grants, rebates, or tax incentives are available through local governments and utilities. The site is updated each week and features new programs as well as changes to existing ones.

Home owners can also finance the purchase and installation of renewable energy systems through home-equity loans. This strategy can help bring down costs through tax savings, since interest payments on mortgage loans are tax deductible.

CONTACTS: Connecticut Clean Energy Fund, www.ctinnovations.com; Database of State Incentives for Renewable Energy, www.dsireusa.org; Interstate Renewable Energy Council, www.irecusa.org.

6
GREEN THREADS

Organic Undies and Recycled Sneaks

It's a lot easier to worry about what we put in our bodies than what we put *on* them. But while organic crops are taking over the produce sections of most supermarkets, organic clothing is making inroads only very slowly. It's not just that people don't want to wear bamboo T-shirts (which are unbelievably soft, like that worn-to-a-whisper T-shirt you stole from your boyfriend and sleep in every night), it's the remaining stigma surrounding the prefix "eco" paired with the word "fashion." While there are still some scratchy-looking hippie clothes out there in the organic clothing world, eco-fashion has made huge strides with catwalk-worthy threads that are anything but crunchy. And even while eco-clothes have gone upscale through such companies as EDUN (pop star Bono and wife's high-end jeans brand), they've also gained wider accessibility with organic lines popping up everywhere from American Apparel to Wal-Mart. And the accessories market has never been hotter, with one-of-a-kind jewelry and handbags that turn everything from skateboards to tires into wearable art.

Are the materials used in athletic shoes environmentally harmful, and can old shoes be recycled?

Margaret Southgate, Hamilton, New Zealand

The ingredient that gives some athletic shoes their cushioning support is sulfur hexafluoride, known as SF_6. It's a popular man-made gas with a uniquely buoyant chemical structure. Unfortunately, SF_6 is also an unusually persistent global warming gas that is more damaging to the atmosphere (molecule by molecule) than CO_2.

Nike's "Air" technology used 288 tons of SF_6 a year, accounting for 1 percent of worldwide production, before they began to phase out SF_6 use in the mid-1990s. According to a spokeswoman from the Nike Environmental Action Team, the company found out that SF_6 was environmentally damaging in 1992. And it wasn't a small effect, either. At the peak of SF_6 production in 1997, Nike sneakers had a greenhouse effect equivalent to more than a million tailpipes. The company began investigating alternative materials.

In 2001, Nike partnered with the Center for Energy & Climate Solutions and the World Wildlife Fund, making a commitment to complete the phaseout of SF_6 by June 2003. That first goal proved elusive. "Most of the transition has happened, but we've run into complications in some of our newer and more technical products in terms of finding a suitable substitute," said Veda Manager, director of global issues management at Nike, around the time of the deadline.

Nike set a new goal to end SF_6 use by 2006, and this one it kept. The new Air cushion is provided by nitrogen, making possible the $160 Air Max 360 (the first sneaker to cushion its entire sole on a bed of air). It was a hit too! The green solution proved to be win-win for the company.

There are other environmental issues with shoes, when you consider the resources and energy that go into making our feet comfortable. And athletic shoes aren't a long-term asset. Perhaps in exchange for its overuse of SF_6, Nike is making an attempt to reduce running shoe waste. The company will take back its shoes, and other brands, grind them up and reuse them in athletic surfaces–granulated rubber from the shoe outsole becomes artificial soccer, football, and baseball field surfaces and weight-room flooring; granulated foam from the midsoles becomes synthetic basketball courts, tennis courts, and playground surfacing tiles; and fabric from the shoe uppers becomes the padding under hardwood basketball floors.

Nike's Reuse-a-Shoe celebrated its thirteenth anniversary in 2008; it recycles between one and two million pairs of postconsumer and defective shoes every year. There's also the option of donating sneakers to local charities and thrift stores. The Children's Rights Foundation (CRF), for one, sponsors an annual used athletic shoe drive through different retail shoe shops nationally. Retailers promote the shoe drive through normal means of advertising. Customers are directed to bring their used wearable shoes to participating stores in exchange for a discount on new shoes as decided by individual retailers. CRF then donates the used sneakers to needy and at-risk children and their families within the United States and abroad.

Some local recycling services will also take your old wearable sneakers and shoes and direct them to those in need. One such service is Eco-Cycle, a nonprofit recycler based in Boulder, Colorado. The organization's Center for Hard-to-Recycle Materials program will take your old pairs of shoes—as well as accessories and other clothing—and send them to relief agencies in developing countries.

CONTACTS: Center for Energy & Climate Solutions, (703) 379-2713, www.energyandclimate.org; Children's Rights Foundation, www.crfi.org; Eco-Cycle, www.ecocycle.org; Nike Environmental Action Team, (503) 671-8044, www.nike.com.

I've found environmentally friendly shoes for adults but have had trouble finding similar shoes for my kids. Are they out there?

Dawn Masterson, Augusta, GA

Kids' shoes are a quickly expanding market filled with mini versions of everything from flip-flops to slippers to heeled dress shoes, and companies with a green perspective have jumped into the race. Green kids' shoes from such makers as Simple, which offers organic cotton ecoSNEAKS with car-tire soles, might seem expensive at forty dollars or more, but they are durable enough to get passed around from sibling to sibling or parent to parent. "It is an investment if you're going to do quality," says Craig Throne, general manager of footwear at Patagonia. "But the shoes do get handed around a bit. If you make a quality product, it doesn't get trashed."

Patagonia has been making climbing gear and outdoor wear for over thirty years and is committed to using sustainable materials—including recycled polyester and only organic cotton in its clothes. Using hemp and recycled rubber content, the company has created kids' shoes that are rugged and sturdy enough for hiking or climbing or simply running around the backyard. Of course, packaging plays a big role, and in Patagonia's case that means boxes made of 100 percent recycled content, with soy-based inks and fun graphics that encourage kids to reuse them. "We're getting kids to participate and be more aware of the outdoor world," says Throne.

Timberland has launched its own line of sustainable kids' shoes. "Kids today are learning about the environment at a younger and younger age—in many cases, they're even teaching their parents," says Lisa DeMarkis, head of Timberland's kid's division. "It's important to show kids that even small choices can have a positive impact." The company strives to use the most environmentally friendly materials when possible—like recycled PET soda bottles in linings or meshes, recycled laces, and organic cotton canvas—but always making sure that the shoes meet performance goals first. "At the end of the day, the shoe has to stand up to kids and their daily adventures," DeMarkis says. Curious customers can read the "nutritional labels," which include the amount of renewable energy used in production, right on Timberland's shoe boxes, which are made from 100 percent recycled postconsumer material.

If you're a parent who wants to avoid leather in your kids' shoes, whether for ethical or environmental reasons, you'll have to do a bit of hunting online. While many vegetarian and vegan clothing sites have yet to add kids' shoes, kidbean.com does, including the popular baby shoes called IsaBooties, which are made with soft, synthetic

Ultrasuede. And for parents of budding dancers, a vegan alternative ballet slipper can be had from the Cynthia King Dance Studio in Brooklyn, New York. The dance instructor and studio owner approached a local shoemaker when she couldn't find an affordable source for vegan slippers and now provides them to the world at large.

CONTACTS: Cynthia King Dance Studio (vegan ballet slippers), www.cynthiakingdance.com; IsaBooties, www.isabooties.com; kidbean.com, www.kidbean.com; Patagonia, www.patagonia.com; Simple, www.simpleshoes.com; Timberland, www.timberland.com.

What's up with these "eco-fashions" I keep hearing about?

Glenn Hammond, San Francisco, CA

What we're talking about here is stylish clothing that uses environmentally sensitive fabrics and responsible production techniques.

The nonprofit Sustainable Technology Education Project (STEP) defines eco-fashions as clothes "that take into account the environment, the health of consumers, and the working conditions of people in the fashion industry." The clothes and accessories that can get over those hurdles are usually made using organic raw materials, such as cotton grown without pesticides, or reused materials such as recycled plastic from old soda bottles. Eco-fashions don't involve the use of harmful chemicals and bleaches to color fabrics and are made by people earning fair wages in healthy working conditions.

Designers have been playing around with organic and natural fibers for years, but so-called eco-fashions had their coming-out party during New York City's Fashion Week in 2005 when the nonprofit Earth Pledge teamed with upscale clothing retailer Barneys New York to sponsor a special runway event called FutureFashion. At the event, famous and up-and-coming designers showcased outfits made from eco-friendly fabrics and materials including hemp, recycled poly, and bamboo. Barneys was so enthused that it featured some of the environmentally sensitive designs in its window displays for several weeks following the event, imparting a unique mystique to this emerging green subset of the fashion world.

One of the highlights of FutureFashion was a stunning pink-and-yellow skirt made from corn fiber by übercool Heatherette designer Richie Rich. "It's definitely something we're going to continue toying with," Rich told reporters. "People often perceive the fashion world as superficial, so it's great to work with materials that are actually good for the environment. I had my doubts, but when we actually saw the fabric swatches we were blown away. They were gorgeous, and it wasn't hard to design with them."

The party moved to the West Coast in June when San Francisco culminated its World Environment Day celebration with a Catwalk on the Wild Side, an eco-chic fashion show sponsored by the non-profit Wildlife Works, featuring top models and designs from the likes of Ecoganik, Loomstate, and Fabuloid.

One of the pioneers of the emerging eco-fashion movement is designer Linda Loudermilk. Her "luxury eco" line of clothing and accessories uses sustainably produced materials made from exotic plants including bamboo, SeaCell (seaweed and cellulose), soy, and *sasawashi*. The latter is a linenlike fabric made from a Japanese leaf that contains antiallergenic and antibacterial properties. Loudermilk also incorporates natural themes in each season's line—her most recent one being an ocean motif. "We aim to give eco glamour legs, a fabulous look, and a slammin' attitude that stops traffic and shouts the message: eco can be edgy, loud, fun, playful, feminine (or not), and hyper-cool," Loudermilk says.

CONTACTS: Earth Pledge, www.earthpledge.org; Greenloop, www.thegreenloop.com; Heatherette, www.heatherette.com; Linda Loudermilk, www.lindaloudermilk.com; Sustainable Technology Education Project, www.stepin.org; Wildlife Works, www.wildlifeworks.com.

I've been hearing a lot about all the recycled materials being turned into handbags and purses. Are these bags actually fashionable?

Mary-Beth Johnstone, Cos Cob, CT

Eco-fashion, especially in the world of bags, purses, and carriers, has proven to be an inventive outlet for all kinds of recycled material. And yes, most of these bags—even those made from such unlikely materials as candy wrappers (by Ecoist) or carpets (CarpetBags)—not only look good but would probably draw looks of admiration from fellow bag aficionados.

The Canadian site eco-handbags.ca carries an unbelievable assortment of creatively adapted materials turned to wearable art by green handbag companies. There are bags made from old books, boat sails, juice boxes, aluminum cans, plastic bottles, neckties, cigar boxes, skateboards, chopsticks, soda pop tops, and bicycle tire inner tubes. And these don't look like they've been knit together from a trash bin—they're impeccably sewn, one-of-a-kind accessories. The juice box cooler bag, handmade by a cooperative in the Philippines for the company Bazura Bags, would make a great general-purpose carryall, while the sleek Roadster Handbag made of truck tire inner tubes by English Retreads would make a stylish everyday purse.

Ava DeMarco and her husband, Rob Brandegee, looked at used license plates one day and saw handbags. The couple launched their company, Little Earth Productions, in 1993 with a mission to match style with eco-consciousness. At first, license plates were used as ornaments on recycled rubber bags. Then they became the bags themselves, twisted into colorful cylindrical purses. Now Little Earth's recycled license plate handbags can be found in more than one thousand retail outlets and in the clutches of everyone from Oprah to Chelsea Clinton. "Everything we make is one of a kind, because all license plates are unique," says DeMarco. In one year, Little Earth actively recycled more than fifteen tons of rubber and forty thousand license plates.

And why not turn all that old tire rubber into something eminently wearable? The material is completely durable and hardy and effective for everything from men's messenger bags to women's clutches. "I've always been aware of the tire situation," says Robin Gilson, president and founder of Vulcana, a company that makes bags out of recycled car tires. "They collect water; they are breeding grounds for mosquitoes. I thought: 'Wouldn't it be great if you could melt car tires down and reshape them?' " After taking a leave of absence from her job as an attorney in 1995, Gilson tracked down a company that would take recycled car tire crumb and mix it with natural rubber to create a material suitable for stitching into bags. Vulcana launched its product line in 2001. The company takes 30 to 50 percent of its material from recycled car tires. The rest is virgin rubber, mostly from small, family-owned plantations in Malaysia. Some products are hemp-fused, which means the rubber is cured directly onto a hemp fabric.

For animal-conscious environmentalists, the new range of handbags has been especially welcome—whether they're made from tires, records, hemp, or chopsticks, these bags are a great alternative to leather and an easy way to make a fashion statement.

CONTACTS: Bazura Bags, www.bazurabags.com; Eco Handbags, www.eco-handbags.ca; Ecoist, www.ecoist.com; English Retreads, www.englishretreads.com; Little Earth Productions, www.littleearth.com; Vulcana, www.vulcanabags.com.

Where can I find fashionable clothing brands that use organic materials?

Trey Muhlhauser, Chicago, IL

The fashion industry is nothing if not trendy, so it's jumped on the eco-fabric bandwagon big time. Organic materials are being incorporated into designs from the runways of Milan to the aisles of Kmart. Materials like hemp and bamboo are coming on strong, but organic cotton is by far the fabric of choice for most green clothing designers. According to Organic Exchange, a nonprofit committed to expanding the use of organically grown fibers, global retail sales of organic cotton products increased from $245 million in 2001 to $583 million in 2005.

The problem with traditional cotton—by far the most popular clothing fabric in the world, amounting to a $300 billion global market—is that producers use liberal amounts of insecticides, herbicides, and synthetic fertilizers to grow it. Analysts estimate that cotton crops use about one-quarter of all the agricultural insecticides applied globally each year. According to the EPA, seven of the top fifteen pesticides used on U.S. cotton crops are potential or known human carcinogens.

Given such problems, choosing organically grown alternatives may be one of the best things consumers can do to help the environment. Luckily, many designers are using such materials to great effect in their newest lines. Examples include Kelly B Couture, Xylem, Turk+ Taylor, Blue Canoe, Stewart+Brown, Armour Sans Anguish, Ecoganik, NatureVsFuture, EcoDragon, Gypsy Rose, Maggie's Organics, Two Star Dog, and Enamore, all of which are making waves in fashion circles with their cutting-edge clothing designs crafted from materials grown without harmful synthetic chemicals. Big players like Levi Strauss, Victoria's Secret, Esprit, Patagonia, and Timberland are also increasingly offering organic cotton products.

Singer Bono, along with his wife, Ali Hewson, and designer Rogan Gregory, launched the EDUN brand in 2005, offering organic cotton T-shirts and sweatshirts made in Tunisia and Peru. A key part of EDUN's mission involves fair wages and healthy working conditions for garment workers in developing countries.

Some online retailers featuring hip clothing made from organic materials include upstarts like Envi, Bamboo Styles, Grass Roots Natural Goods, and better-known outlets like Gaiam. Even Wal-Mart and Target are now stocking a wide range of organic cotton clothing. To find other organic clothing retailers, check with *E* magazine's own online Marketplace and the EcoMall, both of which link to a wide range of cool, green-friendly garments. Another website, EcoBusinessLinks, provides a listing as well on its "Natural Clothing Retailers" page.

The nonprofit Organic Consumers Association runs Clothes for a Change, a campaign to pressure major clothing retailers and manufacturers to wean themselves off traditional cotton and petroleum-derived polyesters and to start using more organic materials. Another key element of the campaign is to educate consumers about the benefits of clothing made from organic materials.

CONTACTS: *E* magazine Marketplace, www.emagazine.com/marketplace; EcoBusinessLinks, www.ecobusinesslinks.com/natural_clothing_natural_fibre_clothes.htm; EcoMall, www.ecomall.com/biz/clothing.htm; EDUN, www.edunonline.com; Gaiam, www.gaiam.com; Organic Consumers Association's Clothes for a Change Campaign, www.organicconsumers.org/clothes.

I've heard that cotton is more environmentally friendly than synthetic fabrics. But what is the ecological impact of cotton?

Christina Wong, Salt Lake City, UT

Cotton is as natural as can be, and it's certainly more biodegradable (and less flammable) than polyester fabrics, but unfortunately conventional growing practices literally soak it in toxins. The Organic Trade Association (OTA) says the cotton industry uses approximately 25 percent of the world's insecticides and more than 10 percent of its pesticides. Producing the cotton fabric required for a regular T-shirt releases a quarter pound of chemical fertilizers and pesticides into the water, air, and soil.

According to the OTA, the EPA considers seven of the top fifteen pesticides used on cotton in 2000 in the United States as "possible, likely, probable, or known human carcinogens." With increased consumer awareness, the demand for organically grown cotton continues to rise. In 1994, the Sustainable Cotton Project began an effort to network farmers, manufacturers, and consumers to pioneer markets for certified organically grown cotton. Numerous retail companies specialize in organic cotton, including bedding manufacturer Coyuchi, outdoor clothing guru Patagonia, and furniture maker H3Environmental.

LaRhea Pepper of the Texas Organic Cotton Marketing Cooperative says, "Most consumers are concerned about what they put in their bodies, but the same awakening hasn't happened for what they put *on* their bodies. If you have industrial pressed sheets, you could be sleeping on chemicals."

CONTACTS: Coyuchi, (888) 418.8847, www.coyuchiorganic.com; H3Environmental, (818) 766-1787, www.h3environmental.com; Organic Trade Association, (413) 774-7511, www.ota.com; Patagonia, (800) 638-6464, www.patagonia.com; Sustainable Cotton Project, www.sustainablecotton.org.

What impact does mining for diamonds and other gems have on the environment?

Tiffany Schultz, Dayton, OH

The glitter in the showroom windows doesn't tell the whole story. Gem mining can be very destructive, causing soil erosion and sedimentation, water pollution and depletion, poisoning of wildlife and vegetation, flooding and even landslides. The contents of "mine tailings"—rock and other waste materials separated and left behind in the mining process—wreak havoc on nearby agricultural lands, and pose myriad human health problems.

In the United States, mining companies are legally obligated to conduct environmental impact studies of proposed sites and then, if approved by regulators, follow the letter of the law regarding the protection of wildlife, air, and water and the proper disposal of hazardous waste. Furthermore, many U.S. states have "reclamation" laws on the books calling for the safeguarding of surface and groundwater around mining operations and cleanup and revegetation after the fact to restore mining areas to their original condition.

But mines outside U.S. borders are not subject to the same rules, even if run by American companies. Large-scale demand means large-scale mining operations, and that often means massive amounts of sedimentation and tailings falling into water systems around the

world. The mercury and cyanide used to separate gold and copper from rock also make their way into our air and water.

With no country-of-origin labeling laws or system in the jewelry and gem trade, consumers can never be sure if their bracelets, rings, and necklaces come from responsible sources or from companies whose mining operations are polluting, destroying wildlife habitat, exploiting poor or indigenous people (and their resources), or funding a civil war, as does the diamond trade in Angola and Sierra Leone.

According to Friends of the Earth, in 1996 mining giant Freeport-McMoRan Copper & Gold was dumping 110,000 tons of mine tailings into the local river system on a daily basis. Plans to expand Freeport's mining activities in Indonesia, according to the company's own environmental auditors, could "increase its dumping of untreated tailings to 285,000 tons daily," presenting serious health challenges for local residents who have little in the way of power or resources to halt such activity.

But buying jewels doesn't have to aid and abet environmental destruction. People have been finding valuable gems and minerals for centuries by panning for them themselves in rivers and streams. There are even "theme parks" scattered across the United States, such as Gold City Gem Mine in Franklin, North Carolina, that let you "mine your own gemstones."

And companies such as Junk to Jewels and Snooty Jewelry sell "gems" made from recycled materials, handmade beads, and glass. Another company, Global Marketplace, sells a wide range of jewelry made by artists in developing countries such as Nepal, Mexico, and Chile, thus helping producers in these nations increase their standard of living above the poverty line.

CONTACTS: Friends of the Earth, (877) 843-8687, www.foe.org; Global Marketplace, www.globalmarketplace.org; Gold City Gem Mine, (800) 713-7767, www.goldcityamusement.com; Junk to Jewels, www.junktojewels.net; Snooty Jewelry, www.snootyjewelry.com.

Are there any good alternatives to leather, and why should we worry about it?

Brianna Jacobs, Somerville, MA

Leather is everywhere—from shoes and belts to purses, wallets, jackets, and furniture. Most people probably assume the leather that finds its way into their wardrobes is a by-product of the meat industry. In truth, most of our leather is sourced from countries like China and India, which don't have strict animal welfare laws. And while cows are certainly the animals used most commonly for leather goods, a host of animals may end up in our bags and belts, including horses, deer, sheep, and in more exotic cases alligators or snakes. All of which may make an animal lover or vegetarian queasy.

But environmentalists have reason to forgo leather too. Processing leather requires copious amounts of energy and a toxic stew of chemicals including formaldehyde, coal tar, and cyanide-containing finishes. The tanning process is just as laced with pollutants and can leave chemicals in the water supply (as described in the bestselling book *A Civil Action*) and on the hands and in the lungs of workers in the developing world.

Tanneries are top polluters on the EPA Superfund list, which identifies the most critical environmental cleanup sites. The website organicleather.com reports that due to their toxicity "many old tannery sites can't be used for agriculture, or built on, or even sold." That website is run by Mill Valley, California, retailer Organic Leather, which offers a return to the tanning practices of old—using animals that are organically fed and humanely raised and a tanning process that employs plant tannins, vegetable tannins, or smoke to cure the leather, with zero toxicity in the process.

But with the wealth of fashionable faux-leather alternatives, there's no need to ever wear skins. Cruelty-free fashions have advanced in leaps and bounds, with variations on every style of handbag, wallet, belt, and boot. The vegan boutique Alternative Outfitters even has a version of the ubiquitous UGG boot made with microsuede "shearling" on the outside and synthetic wool inside, while Iowa company Heartland Products, Ltd. sells Western-style nonleather boots and nonleather Birkenstock sandals. Science has come up with plenty of comfortable, durable alternatives to materials made with animal products. These include vegan microfiber, which claims to match leather in strength and durability, and pleather, Durabuck, and Nusuede.

The products made with these synthetic materials tend to be less expensive than their leather counterparts and are being produced by major manufacturers like Nike, whose Durabuck athletic and hiking shoes "will stretch around the foot with the same 'give' as leather . . . and are machine washable," according to company sources. The People for the Ethical Treatment of Animals (PETA) website includes an alphabetized "Cruelty-Free Clothing Guide" that is teeming with resources for nonleather bags, belts, wallets, shoes, jackets, pants, and many other items. And you won't need to adjust your style, either. Websites like Vegetarian Shoes and Bags offer everything from purple faux-snakeskin peep-toe pumps for hitting the clubs to hemp sneakers with recycled outsoles that look skate park ready and distinctive pleather bags and versatile woven belts.

CONTACTS: Alternative Outfitters, www.alternativeoutfitters.com; Heartland Products, Ltd., (800) 441-4692, www.trvnet.net/~hrtlndp; Organic Leather, (888) 886-8959, www.organicleather.com; People for the Ethical Treatment of Animals' leather site, www.cowsarecool.com; Vegetarian Shoes and Bags, www.vegetarianshoesandbags.com.

7
THE WHOLE KID AND CABOODLE

From Dirty Diapers to Toxic Toys

Wouldn't it be great if the world were one safe, breathable place where kids could play sports without worrying about pesticides in the playing fields, sit at desks without the danger of inhaling chemical fumes, and eat food that hasn't been processed and sweetened and artificially colored into some unrecognizable brightly colored lump? Yes, dreams are nice. In the real world, however, parents have to be vigilant about their kids' health, beginning when the little tykes are still in utero. Pregnant moms have to watch out for mercury in formerly healthy products like tuna fish and chemical phthalates in perfectly innocent-looking hand lotions. Then they have to debate the benefits of which diapers to clad their baby in. And when parents are done fretting through the toddler years, they have to get active at the local elementary school to make sure the essentials of environmental education are being taught and that there are no idling school buses wreathing the building in fumes. Because someone, after all, has gotta save this planet.

Which are better for the environment, disposable or cloth diapers?

Barbara Fritts, White Lake, MI

The "disposable versus cloth" debate has raged for years. Nondegradable disposable diapers can sit for decades, even centuries, in landfills, and they require thousands of tons of plastic and hundreds of thousands of trees to manufacture. But the water and chemicals used to clean cloth diapers, and the fossil fuels diaper services consume to transport them, suggest that their relative environmental impact could be a wash.

Still, the environmental balance may be shifting back to cloth diapers, which have been slowly losing what little market share they had. Modern advances in water and energy efficiency in washing machines and dryers have reduced the environmental impact of diaper laundering. Much as they'd probably want to avoid it, concerned parents also have to think about, well, sewage. The urine and feces in disposable diapers enter landfills untreated, possibly contaminating the groundwater supply. Whether cloth diaper waste is flushed down the toilet or removed in the washing machine, that dirty water will enter a sewer system and, most likely, a wastewater-treatment plant.

John Shiffert, executive director of the National Association of Diaper Services (NADS), points out that the chlorine by-product dioxin, a carcinogen, has been found in trace amounts in disposables.

If you're concerned about the environment and want the convenience of disposables, you can try Nature Boy & Girl's competitively priced, cornstarch-based diapers, which can be composted. Using flushable cloth diaper liners, made by Tiny Tush and other companies, means only the thinnest—and messiest—part gets thrown away. Parents who want to use cloth diapers can hire a cleaning service to do the dirty work. Their numbers have rebounded in recent years. Check the Yellow Pages, or contact NADS to locate a service in your area.

CONTACT: National Association of Diaper Services, (610) 971-4850, www.diapernet.com; Nature Boy & Girl, (206) 784-7766, www.natureboyandgirl.net; Tiny Tush, (608) 356-2500, www.tinytush.com.

Is it true that toxins in some common childhood vaccines cause autism, and if so should I not have my children vaccinated?

Peter Fox, Brewer, ME

The concern centers on thimerosal, a mercury-based preservative once common in vaccines, as a potential culprit in the rise of autism in recent years. Preservatives like thimerosal are used to prevent infection in the event that a dose is accidentally contaminated. Due to recent heightened concerns over the potential effects of mercury on child brain development, though, most vaccines for U.S. children under the age of six no longer contain thimerosal.

The issue received considerable attention following an article by environmental lawyer and activist Robert F. Kennedy Jr., who claimed that federal officials covered up proven scientific links between thimerosal and a fifteenfold increase in autism since 1991. At that time, the FDA and the Centers for Disease Control had recommended that three additional vaccines containing thimerosal be given to infants.

"More than 500,000 kids currently suffer from autism, and pediatricians diagnose more than 40,000 new cases every year," says Kennedy. "The disease was unknown until 1943, when it was identified and diagnosed among 11 children born in the months after thimerosal was first added to baby vaccines."

Vaccine manufacturers have begun to phase thimerosal out of injections given to American infants. Unfortunately, though, according to Kennedy they have continued to export their back stock of tainted vaccines to developing countries. Autism was virtually unknown in China prior to the introduction of thimerosal by U.S. drugmakers in 1999; today approximately 1.8 million Chinese children suffer from the disorder. Even so, industry groups complain that a direct link between autism and thimerosal has not been definitively proven.

To be safe, parents may want to ask their pediatrician if the vaccines he or she uses contain thimerosal. Some flu and tetanus shots with thimerosal are still given to preteens in the United States, although preservative-free versions are usually available upon request. On its website, the FDA provides a listing of common children's vaccines and their thimerosal content, if any, and also lists thimerosal-free alternatives.

Parents considering not vaccinating their children should know that this is a hotly debated topic, and *E* is in no position to recommend a course of action. Most medical professionals argue that vaccines have saved more lives than any other kind of medical intervention and recommend their use to guard against polio, diphtheria, rubella, hepatitis B, and many other diseaseas. But some critics believe that the medical benefits of vaccines are exaggerated and that negative reactions to toxic chemical ingredients in many vaccines have been grossly underreported.

CONTACTS: Centers for Disease Control and Prevention "Mercury and Vaccines" page, www.cdc.gov/od/science/iso/concerns/thimerosal.htm; Food and Drug Administration, "Thimerosal in Vaccines" information, www.fda.gov/cber/vaccine/thimerosal.htm; Wikipedia "Vaccines" page, http://en.wikipedia.org/wiki/vaccine.

Some art supplies, like glues and markers, can be quite toxic, especially to children. Are there eco- and health-friendly alternatives out there?

Frances Goulart, Austin, TX

Look for a seal of approval from the Art & Creative Materials Institute, which certifies that a product "contains no material in sufficient quantities to be toxic or injurious to humans, including children, or to cause acute or chronic health problems."

Since 1988, federal mandates require that all art materials be reviewed for toxicity, and the federal government has long required labeling of acutely toxic art supplies (those which could cause immediate harm). But deciphering the codes is quite a task. Products found to pose a chronic health threat are clearly labeled, but nonhazardous materials are marked "conforms to ASTM D-4236." That label does not mean the product is nontoxic, but that it has been evaluated by a toxicologist and identifies hazardous materials and safe-use instructions.

A number of companies also provide environmentally friendly art materials. Eco-House's paints and wood finishes are all free of aromatic and chlorinated hydrocarbons, which are potentially problematic. D'UVA Fine Artists Materials produces powder-based acrylics that use heat as a fixative, replacing the need for conventional aerosol-based spray fixatives, which often rely on flammable, potentially toxic, and ozone-destroying chemicals.

CONTACT: Art & Creative Materials Institute, (781) 293-4100, www.acminet.org; Eco-House, (506) 366-3529, www.eco-house.com.

Does environmental education figure prominently in classrooms these days? By that I mean not just science but an understanding of key issues and environmental stewardship.

Mary Swan, Framingham, MA

Environmental education has long struggled for legitimacy, even though studies in the late 1980s revealed that schoolchildren lacked basic knowledge about the natural world. In 1990, Congress passed the National Environmental Education Act, forcing the EPA to strengthen and expand environmental education nationwide through education and teacher training and the administration of grants to exemplary programs.

The EPA launched some good programs, but a lack of funding has prevented many ideas from moving forward. Between 1991 and 1996, the EPA received ten thousand environmental education grant applications, totaling $300 million, but was only able to fund twelve hundred, totaling $13 million. Further cutbacks occurred throughout the Bush administration.

Responsibility for teaching kids about the environment has fallen on local schools and individual teachers. According to the President's Council on Sustainable Development, because environmental education is multidisciplinary, it is hard for teachers to work it into their narrowly defined lesson plans. Also, most teachers are not trained in environmental subjects. As a result, nongovernmental organizations have become increasingly involved with environmental education efforts.

One such organization is the North American Association for Environmental Education (NAAEE), a network of volunteers that provides guidelines and resources for educators and parents who want environmental education for their K–12 students. According to NAAEE's Mary Ocwieja, the group takes a "cooperative, nonconfrontational, and scientifically balanced approach" to education about environmental issues.

Another organization is the National Environmental Education Foundation (NEEF), chartered by Congress in 1990, which sponsors Classroom Earth, a free website that calls itself "the best of the best" collection of environmental education programs and resources for K–12 teachers, parents, and students. The site helps educators, afterschool programs, and home-schooling parents find up-to-date information on the most successful, well-tested, and effective national environmental education programs available today.

All this work is starting to pay off. Some 61 percent of U.S. K–12 teachers say they include environmental topics in their curriculum, with some devoting hundreds of hours of classroom time annually to environmental issues.

CONTACTS: Classroom Earth, www.classroomearth.org; North American Association for Enviromental Education, www.naaee.org.

I would like my children to start eating organic foods. Are there any organic products that young kids would enjoy?

Amanda Seth, Rockland, ID

Each day, more than a million children ages five and under take in unsafe levels of pesticides from food consumed at home.

"Infants and children are especially vulnerable to the harmful effects of pesticides, which can include cancer and nerve damage. Typically, the younger a child is, the greater the degree of susceptibility," says environmental health researcher Linda Bonvie. "Each exposure to a toxic chemical adds to a kid's body burden, and since children can't detoxify as well as adults can, they need to be protected from pesticides and environmental poisons wherever possible," she adds.

Fortunately, organic foods designed especially for kids are turning up more and more in home kitchens and on school lunch trays. Whole Foods Market, for instance, offers an entire line of organic foods for kids. "Many of our shoppers wanted to provide kids with organic food choices, but a lot of traditional foods didn't appeal to a kid's palette," says Whole Foods brand manager Linda Boardman. Whole Kids products, including organic peanut butter, string cheese, and flavored applesauce, are available at Whole Foods stores in twenty-eight states as well as in Canada and the United Kingdom.

Meanwhile, other national natural foods grocery chains such as Trader Joe's offer a wide range of foods safe for kids to eat, including organic juices from R.W. Knudsen and Santa Cruz Organic in flavors ranging from apple to tropical. Stonyfield Farm's YoBaby organic yogurt is available for babies and toddlers, and the company's colorful tubes of YoSqueeze are designed to be easily packed in a lunchbox.

It's also best to avoid breakfast cereals laden with sugar and preservatives. One good alternative is any of the cereals—including Amazon Frosted Flakes, Gorilla Munch, Koala Crisp, and Orangutan-O's—made by EnviroKidz, which uses less sugar and all organic ingredients in its products.

For snacks, kids can also choose from a wide range of organic choices. Planet Harmony offers organic jelly beans, fruit snacks, and gummy worms, and Country Choice Naturals sells organic animal cookies. And parents shouldn't forget that fruit from an organic farmers' market can satisfy hungry kids.

For health-conscious parents, it's become a whole lot easier to avoid pesticides, preservatives, and sugars and still give kids food they'll eat.

CONTACTS: Country Choice Naturals; www.countrychoicenaturals.com; EnviroKidz, www.envirokidz.com; Planet Harmony; www.harmonyfoods.com; R.W. Knudsen, www.knudsenjuices.com; Santa Cruz Organic, www.scojuice.com; Stonyfield Farm, www.stonyfield.com; Trader Joe's, www.traderjoes.com; Whole Foods Market, www.wholefoods.com.

The marketing of soda to schoolkids was a big item in the news this past year. What's so bad about soda, and where can I find healthier alternatives that still have the "fizz"?

Chase Abromovitch, via e-mail

The United States is the land of the soft drink. We've got 450 different types to choose from and more than 2.5 million vending machines dispensing them around the clock, even in our schools. In 2004, 28 percent of all beverages consumed in the United States were carbonated soft drinks. And we've got the expanding waistlines to prove it.

The Department of Agriculture advises a two-thousand-calorie-a-day limit as part of a healthy lifestyle, and no more than ten to twelve teaspoons of sugar. But since 1990, the amount of sugar in American diets has increased by 28 percent, with about a third of it coming from soft drinks. A single twelve-ounce can of soda has around thirteen teaspoons of sugar, usually in the form of high-fructose corn syrup.

That sweetening syrup causes weight gain by interfering with the body's natural ability to suppress hunger feelings. Currently, 64.5 percent of adults over the age of twenty are overweight, 30.5 percent are obese, and 4.7 percent are severely obese. According to Dr. Sonia Caprio, a Yale University professor of pediatric endocrinology, "The reality is that obesity and type 2 diabetes [are linked] with the consumption of sodas."

In response, the nation's largest beverage makers—including Dr Pepper Snapple Group (formerly CadburySchweppes), Coke, and Pepsi—agreed in 2006 to halt nearly all soda sales in public schools. Beginning in 2009, elementary and middle schools will sell only water and juice (with no added sweeteners), plus fat-free and low-fat milk. High schools will sell water, juice, sports drinks, and diet soda. Diet sodas use artificial sweeteners, which add little or no calories, though some, such as aspartame, are embroiled in controversy over their questionable health benefits and even possible links to cancer.

Michael Jacobson, executive director of the Center for Science in the Public Interest (CSPI), lauds the move. "Soft drink companies have been marketing what we call liquid candy in high schools and some middle schools for many years now," he says. "It will be great to get rid of them."

If you can't do without soda pop, there's no shortage of natural varieties. Many are sweetened with cane juice, because it is less processed and has many of the same nutrients found in sugarcane. Others add no sweetener and instead let the real fruit ingredients do the job. Popular brands include Steaz, a less-carbonated but flavorful drink available in eight flavors; R.W. Knudsen fruit spritzers, which contain only sparkling water and natural flavors and juices and come in sixteen flavors; Santa Cruz Organic sodas, which taste like fresh fruit juice with light carbonation and are made with organic ingredients in ten flavors; IZZE, which offers seven flavors that contain 100 percent pure fruit juice and sparkling water; and WaNu beverages, which taste like slightly less carbonated mainstream sodas.

CONTACTS: American Beverage Association, www.ameribev.org; Center for Science in the Public Interest; www.cspinet.org/new/200605031.html.

Are there prepared lunches comparable to Oscar Mayer's Lunchables that are healthier and more environmentally friendly? And what about that packaging waste?

Carla Bahun, Marietta, GA

Oscar Mayer Lunchables appeal directly to kids with a combination of bright packaging, fun-to-eat snacks, and cross-promotions with blockbuster movies. But inside those yellow boxes lies a spectrum of potential health disasters. The packaged meals derive two-thirds of their calories from fat and sugar.

The Center for Science in the Public Interest listed Lunchables on its list of the top ten foods to avoid because "it would be hard to invent a worse food than these combos of heavily processed meat, artery-clogging cheese, and mostly-white-flour crackers. The regular (non-low-fat) line averages 5½ teaspoons of fat (that's 60 percent of calories) and 1,734 milligrams of sodium."

Lunchables' form of attractive packaging is also environmentally unfriendly. It consists of a plastic tray cut into various compartments, which is then sealed with a transparent and flexible film. This tray is then placed in an outer cardboard box. All this makes it very difficult to recycle, so much so that Lunchables earned a Lifetime Wastemaker Achievement Award from the Massachusetts Public Interest Research Group.

So how are parents to "do the right thing"? Enter Amy Hemmert and Tammy Pelstring, two California moms who were appalled to learn that a typical American schoolkid generates sixty-seven pounds of discarded lunchbox packaging waste per school year. That's more than eighteen thousand pounds yearly for the average-size elementary school!

Hemmert and Pelstring began networking with other parents who shared their concerns and quickly learned that by using reusable lunch containers, cloth napkins, stainless-steel forks and spoons, and refillable drink containers, they could eliminate their kids' lunch waste altogether. It goes without saying that Lunchables and other unhealthy prepared meals were also off their shopping lists.

They also discovered they could save money, as the costs of single-use disposables like juice boxes adds up quickly when compared to doling juice into plastic screw-top "sippie" cups from half-gallon containers. Sure, some of the silverware and containers never make it back home, but that's a small "one step back" against the "two steps forward" of saving hundreds of dollars per child per school year.

Waste-free lunches also save schools time and money, as less waste cuts down on the frequency of trips to the outside Dumpster and on the amount of trash that needs to be hauled away. "If every American child attending a public elementary school packed a waste-free lunch, 1.2 billion pounds of lunch waste would be diverted from landfills each year," says Hemmert. "Landfills would last longer, and children would learn the importance of protecting the planet," she adds.

In 2002, the partners launched a company, Obentec, specializing in the production of stylish reusable and modular lunch containers called Laptop Lunches, fashioned after Asian bento boxes. The company also produces a free monthly newsletter, the *Laptop Lunch Times*, which includes lunch recipe suggestions, packing tips, and links to related websites.

The demand for healthier lunches is creating entrepreneurial opportunity. Brown Bag Naturals, launched by a former investment banker, is currently catering five-dollar fresh lunches to fifteen schools in the Los Angeles area, but it has plans to go national in four cities. Founder Adam Zauder says the natural food companies he worked with had grown frustrated by their inability to get their products into the closed shop known as the school cafeteria.

At schools supplied by Brown Bag Naturals, students might tuck into a vegetable teriyaki bowl with edamame and pineapple chunks one day, and on another a bean burrito, baby carrots, and vanilla yogurt with granola. "These are wholesome versions of the food kids love," Zauder says.

Of course, you can also pack your kids' own healthy, environmentally friendly lunches and snacks. If your kids insist on Lunchable-style meals, a simple alternative would be to create healthy, low-fat snack replacements. A healthier version might include low-fat string cheese, adding vegetables to sandwiches, using low-fat crackers and whole-grain bread instead of white bread, including fruit or juice (100 percent juice only), and 1 percent or fat-free milk.

CONTACTS: Brown Bag Naturals, www.brownbagnaturals.com; Center for Science in the Public Interest, (202) 332-9110, www.cspinet.org; Massachusetts Public Interest Research Group, (617) 292-4800, www.masspirg.org; Obentec, www.laptoplunches.com; Waste-Free Lunches, www.wastefreelunches.org.

What's behind the startling explosion in nut allergies among children? Is it changes in the kids, the peanuts, or the processing?

Lynne Whetzel, Ithaca, NY

If your local elementary school has banned peanut butter sandwiches, it's probably for a good reason. The incidence of nut allergies among American and British children has tripled within the last two decades.

Because the phenomenon seems limited to developed countries, some environmentalists believe that pollution and synthetic chemicals might be to blame. An allergic reaction happens when the body's immune system overreacts to a perceived threat, and researchers believe there may be an as-yet undiscovered link between exposure to various chemicals, pollutants, and food additives and an overall rise in immune system disorders.

Parents of children suffering from nut allergies live life constantly checking the ingredients on food labels. Nuts and nut oils are used in an increasingly wide range of processed foods, including many of the chips and cereals kids love. The ubiquity of snack foods makes it difficult for kids to avoid nuts and nut oils, even if they know they are allergic.

Nut allergies can start early in development and usually remain into adulthood. From the second trimester of pregnancy on, the unborn fetus can recognize allergens to which the mother has been exposed and may already begin to develop sensitivities that can lead to allergic reactions following birth.

Pregnant women with a history of allergic reactions can minimize the risk to their children by avoiding certain known allergens, especially tree nuts (cashews, almonds, pecans, and walnuts) and peanuts. Breast-feeding mothers should also avoid foods that contain these allergens, as they can be transmitted to babies via breast milk. Additionally, several leading brands of creams used by mothers to ease discomfort while breast-feeding contain nut oils that can trigger allergic reactions in babies.

Symptoms of nut allergies can range from mild reactions like watery eyes, an itchy throat, or a runny nose to severe reactions like eczema, hives, nausea, and vomiting. In extreme cases, allergic reactions can lead to anaphylaxis, a life-threatening condition hastened by the body's release of toxic amounts of histamine into the bloodstream.

The best source for further information is the Food Allergy & Anaphylaxis Network (FAAN), a nonprofit organization in Fairfax, Virginia (with a related group in England), which raises awareness about and conducts research on food allergies and anaphylaxis.

Some researchers believe, however, that the threat of nut allergies is vastly overstated by FAAN and others. "The rash of fatal food allergies is mostly myth," writes Meredith Broussard in an annotated response to a FAAN ad in *Harper's* magazine, "a cultural hysteria cooked up with a few key ingredients."

CONTACTS: Anaphylaxis Campaign, www.anaphylaxis.org.uk; Food Allergy & Anaphylaxis Network, (800) 929-4040, www.foodallergy.org.

Are my kids breathing in dangerous exhaust fumes by riding the school bus?

Molly Schink, Winnetka, IL

Nothing looks friendlier than that big yellow school bus, but it's not as cuddly as it appears. The more than twenty-four million children who ride the bus every day (an average of ninety minutes in transit) are routinely exposed to harmful diesel exhaust emissions, a witches' brew that includes carbon monoxide, carbon dioxide, sulphur dioxide, formaldehyde, and tiny soot particles.

The EPA classifies diesel emissions as a "likely carcinogen." Diesel emissions are estimated to be responsible for 70 percent of the cancer risk arising from air pollution, according to the California Air Resources Board. Dangers from diesel exhaust can range from respiratory illnesses including asthma and bronchitis to lung cancer and heart disease.

Children are more vulnerable to the effects of diesel exhaust than adults because they breathe more quickly and take more air into their developing lungs. Approximately 390,000 diesel school buses are on U.S. roads today, and a third were made before 1990 when stricter emissions guidelines were first enforced. According to the Natural Resources Defense Council, a child riding inside a school bus may be exposed to as much as four times the amount of toxic diesel fumes as someone riding in a car directly ahead of it.

Diesel particulate filters, which cost around $700 each, can cut tailpipe emissions by a whopping 85 percent. And "closed crankcase filtration systems," which are installed under the hood and filter the discharges that come directly from the engine's crankcase vent, can cut engine soot by nearly 90 percent at a cost of around $7,500 each. Buses can be retrofitted with one or both filters.

Nationwide, school bus emission-reduction programs are underway with the help of the EPA's Clean School Bus USA program. In addition to retrofit projects, the program seeks to replace older buses with new, less-polluting buses and encourage unnecessary school bus idling. Concerned parents can help reduce their children's exposure to diesel emissions from school buses by advocating at town and board of education meetings for the use of new or retrofitted school buses. Also, bus windows should remain open when weather allows, and children are safer sitting nearer the front of the bus, as exhaust tends to accumulate in the back.

CONTACTS: EPA's Clean School Bus USA, www.epa.gov/cleanschoolbus; Northeast Diesel Collaborative, www.northeastdiesel.org; Natural Resources Defense Council, www.nrdc.org/air/transportation/qbus.asp.

What are the leading causes of child mortality around the world, and what can be done about it?

Susan Hale, Oquawka, IL

The statistics are staggering. In the world's poorest countries, more than thirty thousand children under the age of five die each day from preventable causes related to conditions of extreme poverty. Rock star Bono and others recently tried to call attention to this fact in television ads showing well-known celebrities snapping their fingers every three seconds, each snap representing another tragic child death.

A baby girl born in Sub-Saharan Africa today faces a 22 percent risk of death by age fifteen. More than a third of casualties are babies who don't survive their first month. They suffer from low birth weight due to their mothers' poor nutrition, and then lack access to adequate nutrition themselves. The World Health Organization says that poverty-related malnutrition is the key factor in over half of all childhood deaths.

Many children suffer from debilitating infections virtually right out of the womb, and analysts say that casualties could often be prevented if basic sanitation were available. Drinking-water pollution is a leading culprit. In areas that lack proper sanitation and that may have just one water source, supplies can easily become contaminated from bacteria in human waste and garbage. According to United Nations statistics, as many as four billion people—nearly two-thirds of the global population—lack access to safe, clean water.

The Bill & Melinda Gates Foundation has helped by distributing low-cost antibiotics and sterile medical implements. "Some global health problems, like AIDS, have no easy solution—but this isn't one of them," says computer geek turned philanthropist Bill Gates. "The world has an opportunity to stop millions of newborn deaths each year."

Debt and population issues are also among the underlying causes of this global tragedy. Some poor nations pay more in service of international loans than on the health and education of their people. Yielding to pressure from Make Poverty History advocates, leaders of the world's top industrialized nations have agreed to cancel forty billion dollars in debt owed by the world's eighteen poorest countries. But this only covers about a sixth of the debt owed by, for example, African nations.

High birth rates in Africa and the Middle East mean that many poor countries have surpassed their "carrying capacity." This has a profound effect on the environment as well as on human misery. According to Population Action International (PAI), "More than 200 million women in the developing world today wish to delay or end childbearing but do not have access to modern and effective contraceptives." In spite of this, in its eight years in office the Bush administration steadily cut family planning aid to developing countries in the name of preventing abortions. Says PAI, "U.S. leadership and investments in international family planning assistance are critical in order to ensure healthy mothers, healthy pregnancies, and ultimately, healthy families."

CONTACTS: Bill & Melinda Gates Foundation child health program, www.glf.org/globalhealth/pri_diseases/childhealth/default.htm; Make Poverty History Campaign, www.makepovertyhistory.org; Population Action International, www.populationaction.org.

I read somewhere that babies were being born nowadays with a number of man-made chemicals detected in their bloodstreams. This is pretty scary. How can it be?

Sandra McGregor, Portland, OR

Yes, the toxic legacy can start at birth. An investigation by the Environmental Working Group (EWG) found that American babies are born with hundreds of chemical contaminants in their bloodstreams.

"Of the 287 chemicals we detected in umbilical-cord blood, we know that 180 cause cancer in humans or animals, 217 are toxic to the brain and nervous system, and 208 cause birth defects or abnormal development in animal tests," the EWG report says.

In the month leading up to a baby's birth, the umbilical cord pulses with the equivalent of at least three hundred quarts of blood each day, pumped back and forth from the nutrient- and oxygen-rich placenta to the rapidly growing baby cradled in a sac of amniotic fluid. This cord is a lifeline between mother and baby, bearing nutrients that sustain life and propel growth.

Scientists used to think that the placenta shielded cord blood—and the developing baby—from most chemicals and pollutants in the environment. But the results of EWG's study show otherwise. "Now we know that at this critical time when organs, vessels, membranes, and systems are knit together from single cells to finished form in a span of weeks, the umbilical cord carries not only the building blocks of life, but also a steady stream of industrial chemicals, pollutants, and pesticides that cross the placenta as readily as residues from cigarettes and alcohol," says the report.

"These ten newborn babies . . . were born polluted," said Congresswoman Louise Slaughter (D-NY), who is leading the charge in Congress to hold chemical producers accountable. "If ever we had proof that our nation's pollution laws aren't working, it's reading the list of industrial chemicals in the bodies of babies who have not yet lived outside the womb."

Slaughter had similar tests done on her own blood, which she found to contain polychlorinated biphenyls (PCBs) that were banned decades ago as well as chemicals like Teflon that are currently under federal investigation. "I have auto exhaust fumes, flame retardant chemicals, and in all, some 271 harmful substances pulsing through my veins," she said. "That's hardly the picture of health I had hoped for, but I've been living in an industrial society for more than seventy years."

CONTACT: Environmental Working Group, www.ewg.org.

Can the mercury contained in some seafood harm a developing baby?

Midge Wilson, Utica, NY

Sadly, the answer is yes. According to a federal report, one in ten American women of childbearing age is at risk for having a baby born with neurological problems due to mercury exposure—this means at least 375,000 babies a year are at risk.

Mercury—emitted by smokestacks and released from common household products like old thermometers—is a persistent heavy metal that ends up in rivers, lakes, and oceans and accumulates in the tissues of fish and animals, including people. "Just one seventieth of a teaspoon of atmospheric mercury can contaminate a twenty-acre lake for a year," says Michael Bender, executive director of the Vermont-based Mercury Policy Project.

Most states and the EPA have issued advisories about eating fish that may have high levels of mercury in their tissues. The FDA says that women can safely eat twelve ounces per week of cooked fish. A typical serving size of fish is from three to six ounces. However, the FDA advises pregnant and nursing women and women of childbearing age who may become pregnant to not eat shark, swordfish, king mackerel, or tilefish, which contain high levels of methylmercury.

In 2003, the FDA released test results showing that albacore "white" canned tuna has three times the mercury levels found in the "light" tuna. "FDA's tests confirm earlier findings that white tuna has far more mercury than light," says Bender. "Yet inexplicitly, FDA still refuses to warn women and kids to limit canned tuna consumption—as twelve states have already done—even after their own food advisory committee recommended this over a year ago."

CONTACTS: Food and Drug Administration, (888) 463-6332, www.fda.gov; Mercury Policy Project, www.mercurypolicy.org.

How can I reduce the number and amount of toxins my new baby is exposed to?

Beth Stevenson, Leesburg, VA

Fortunately for new parents, there is an expanding universe of organic and all-natural products that can minimize their baby's exposure to chemicals. Feeding your baby organic food is a good start, because you will avoid the heavy-duty pesticides, herbicides, and fertilizers that are sprayed onto or absorbed into conventionally grown foods.

Companies like Earth's Best Baby Food provide parents with a variety of prepackaged organic baby foods. Parents interested in an even more back-to-basics approach can get assistance in the form of books, supplies, and tips from Fresh Baby. The company's Fresh Start Kit includes everything a parent needs—instructions, recipes, and materials—to produce fresh, healthy, homemade baby food. Another eco-benefit: "By feeding children with all-natural alternatives, families don't use and toss scores of baby food jars," says company spokesperson Christina Kerley.

Since babies spend so much time sleeping, toxins in their cribs, mattresses, and bedding are also a concern. Lifekind makes crib mattresses that combine organic cotton with wool (which acts as a natural flame retardant) to prevent tender lungs from inhaling plastic and chemical fumes. For even sweeter dreams, bedding made from 100 percent cotton—without permanent press and flame retardant substances—is the least-toxic alternative.

Last, parents should shun soft plastic and vinyl baby toys. Manufacturers often add chemicals, called phthalates, to plastic toys as a softener. This chemical can leach from the plastic and, since toddlers tend to put objects in their mouths, expose young children to a substance that has been linked to cancer and reproductive harm. For this reason, the use of phthalates in baby and children's toys is outlawed in fifteen European countries and Japan. Hard plastic toys or, better yet, wooden playthings coated with water-based lacquer can be found at your local toy store.

CONTACTS: Children's Health Environmental Coalition, (310) 820-2030, www.checnet.org/improve_main.asp; Earth's Best Baby Food, (800) 434-4246, www.earthsbest.com; Fresh Baby, (866) 403-7374, www.freshbaby.com; Lifekind, (800) 284.4983, www.lifekind.com.

8
FEELING THE HEAT

Tackling Global Warming from the Backyard to the Beach

Global warming is not as sexy or fun to talk about as new organic fashions and high-tech green gadgets. The subject can be daunting, especially when the scientists bring out those complicated charts with technical descriptions of why water is warming and ice melting in remote places like Antarctica (though the photos of polar bears are cute). But while the little lifestyle adjustments we make definitely help, global warming is the major environmental issue of our time and a huge challenge. Not only is it not a remote, far-off problem, but its effects are having a major impact on our everyday lives right now. The increase in global temperatures, primarily as a result of the oil and coal we burn (and the meat we consume!), has been tied to everything from rising tides and melting glaciers to droughts and even catastrophic "super storms" (by raising water temperatures). Global warming is already taking its toll on sensitive species, and opening doors for disease-carrying mosquitoes that are thriving in regions once too cold for them. If you've noticed that winter is disappearing, or that beaches at your favorite summertime getaway are much shorter on sand, then you've already begun to witness global warming's very immediate effects. It gives new importance to staying cool.

What exactly is the greenhouse effect, and how is it a bad thing?

Suzanne Gladstone, Queensland, Australia

The greenhouse effect occurs naturally when heat from the sun enters our atmosphere but cannot escape because it is blocked by water vapor, CO_2, and other airborne elements, causing a warming of the earth. Without a natural greenhouse effect, the average temperature of the earth would be about zero degrees Fahrenheit instead of its present 57 degrees Fahrenheit.

But increasing amounts of pollutants from manufacturing and power plants, agricultural activities, automobiles, and other sources that burn fossil fuels have led to an excessive buildup of "greenhouse gases" such as carbon dioxide, nitrous oxide, and methane in the earth's atmosphere. Scientists believe that this buildup is exaggerating the naturally occurring greenhouse effect and is to blame for the average temperature on Earth rising by more than one degree over the last century.

The Intergovernmental Panel on Climate Change, a UN-based group of climatologists, predicts that Earth's temperature will continue to rise from two to ten degrees Fahrenheit during this century as a result of human industrial activity. According to the Sierra Club, the likely effects of this global warming include the melting of massive icebergs and glaciers, a rise in sea level, accelerated coastal erosion, more (and more severe) hurricanes, the spread of infectious diseases, and widespread species extinction, among other problems.

To address this crisis, 127 countries have agreed on mandatory curbs on greenhouse gas emissions via an international treaty that went into effect in 2005 called the Kyoto Protocol. The treaty was the outcome of a meeting held in Kyoto, Japan, in 1997, and under its auspices the United States was supposed to cut its greenhouse gas emissions by 7 percent in the years between 2008 and 2012. While representing only 4 percent of the world's population, the United States currently accounts for about 25 percent of the earth's greenhouse gas emissions. The Bush administration, however, refused to sign the UN-brokered agreement, arguing that it would harm the U.S. economy.

The Natural Resources Defense Council (NRDC) charged that some of the U.S. government's own studies countered that claim.

"While industry trade associations have published many misleading claims of economic harm," says NRDC, "two comprehensive government analyses have shown that it is possible to reduce greenhouse pollution to levels called for in the Kyoto agreement without harming the U.S. economy."

In the vacuum created by noncompliance, the Bush administration pushed for technological approaches that would remove CO_2 from the atmosphere and store it below ground or underwater. But environmentalists fear that loading massive mounts of CO_2 into the earth and oceans could wreak ecological havoc in other ways and doubt that human-induced global warming can be solved by American ingenuity alone.

CONTACTS: Intergovernmental Panel on Climate Change, www.ipcc.ch; Kyoto Protocol, www.unfccc.int/resource/docs/convkp/kpeng.html; Natural Resources Defense Council, (212) 727-2700, www.nrdc.org; Sierra Club Global Warming Campaign, (415) 977-5500, www.sierraclub.org/globalwarming.

What can I do, as just one individual, to curb global warming?

Karen Cross, via e-mail

It's a big planet, and global warming is a huge problem, but there's plenty for individuals to do, with travel at the top of the list. For starters, air travel burns more fossil fuels per person than any other form of transport. So if you can get where you're going some other way, you reduce your contribution of greenhouse gases significantly—provided, of course, that at least a planeload of others are doing the same.

The other main offender, obviously, is the private automobile. Driving less frequently, carpooling, and using public transport such as buses and rail can take a big bite out of the greenhouse gases and pollution you are personally responsible for. Also, think about all those short car trips you take that could be replaced by a brisk walk or bicycle ride.

When driving is a necessity, though, save fuel by making sure your vehicle is properly tuned and the tires properly inflated. If you are contemplating the purchase of a new car, consider getting a gas-sipping hybrid, which will often come with tax incentives.

At home, you can fight global warming by buying energy-efficient appliances and keeping older ones serviced, as inefficiencies translate into energy waste. And simply turning down your thermostat can make you more planet friendly while also lowering monthly bills. In cold weather, dress warmly and sleep with warm blankets; in warm weather, dress lightly and open the windows to create drafts; when you go out, turn heat and air-conditioning down or off.

Insulating and weather-stripping your house is another great way to reduce energy use. And if your utility offers check-off options for renewable power sources like wind or solar, take them, even if it costs a buck or two–it's a small price to pay for a healthy planet. And plant a few trees in the backyard. Over their lifetimes they'll remove tons of CO_2 from the atmosphere that would otherwise contribute to global warming.

Various climate-related websites, including Carbon Footprint and TerraPass, offer free online "carbon footprint calculators" so individuals can see and even calculate how their actions contribute to global warming. Another website, SafeClimate helps businesses of all sizes take action on climate change.

CONTACTS: Carbon Footprint, www.carbonfootprint.com; EarthSave, www.earthsave.org/globalwarming.htm; SafeClimate, www.safe climate.net; TerraPass, www.terrapass.com.

In what ways is global warming already affecting us in North America?

Tyler Merson, New York, NY

It's official, 2007 was the second-warmest year on record. Ski resorts from Vermont to Switzerland wondered where the winter snow went. Some ski resorts in the Pacific Northwest blame global warming for the warm weather that's been shutting down their seasons even before they begin. Skimobile sales are taking a hit too. "If it doesn't snow, people don't snowmobile," says industry spokesman Ed Klem.

Here are just a few more examples of climate change's real and present impact. For one, the twenty hottest years since U.S. record keeping began in the 1880s have all occurred since 1983, and 2006 is the hottest year on record.

If you like New England's maple syrup, you'll be dismayed to know that producers report seeing global warming's effect on their seasonal harvesting cycles. Farmers are tapping their trees a month earlier than their ancestors did, and some fear that global warming will eventually reduce the trees' ability to produce high-quality sap. "I think the sugar maple industry is on its way out," says University of New Hampshire professor Barrett Rock.

University of Washington professor of atmospheric sciences Cliff Mass reports that less snow has been falling in Washington State for the last twenty years. "Global warming *is* occurring," he concludes. Also in trouble due to declining snow are New England and midwestern resorts, some of which have had to reinvent themselves as summer destinations.

The loss of sandy beaches due to climate-aggravated rise in sea level is also troubling, and the problem is accelerating. The National Science Foundation's "Metro East Coast" report says that beach erosion will likely double by the 2020s, increase from three- to sixfold by the 2050s, and by as much as tenfold by the 2080s. Already, sand loss has led to large beach replenishment efforts by the Army Corps of Engineers.

And keep plenty of calamine lotion on hand. Researchers at Duke University found that some vines, including poison ivy, may thrive exponentially in a warmer climate. Experiments showed that poison ivy growing in a CO_2-rich environment grew about three times larger than normal and produced significantly more urushiol, the allergenic substance in poison ivy that causes rashes.

Another indicator of increased warming is the retreat of glaciers across western North America. This troubling phenomenon is especially noticeable in the Waterton-Glacier park complex on the U.S.-Canada border. Several major glaciers there have shrunk by half or more in recent decades. On the U.S. side of the border, the number of glaciers in Glacier National Park has dropped from 150 in 1850 to 35 today.

Finally, stronger storms in recent years, like Hurricane Katrina, may be partially explained by global warming. Researchers have found that both the intensity and number of category four and five storms have greatly increased in the past thirty-five years and have linked that phenomenon to warming ocean temperatures.

CONTACT: Intergovernmental Panel on Climate Change, www.ipcc.ch.

I've seen those images of polar bears stranded on small islands of ice and heard that some are now dying by drowning. How are other wildlife populations affected by global warming?

Jessie Walters, via e-mail

Most researchers agree that even small changes in temperature are enough to send hundreds if not thousands of already struggling species into extinction unless we can stem the tide of global warming. And time may be of the essence: The journal *Nature* reported in 2003 that 80 percent of fifteen hundred wildlife species sampled are already showing signs of stress from climate change.

Wildlife is definitely feeling the heat. The Wildlife Society, representing many professionals in conservation management, said in 2004 that global warming was affecting many North American species and could cause major shifts in ecosystems. The group concluded that caribou (reindeer), polar bears, migratory songbirds, and other species have already responded to climate change by shifting habitat, altering their breeding patterns, or changing their migration routes.

The key impact is habitat displacement—quick shifts in ecosystems that species have adapted to over millions of years. Ice giving way to water in polar bear habitat is just one example of this. Another, says the *Washington Post*, is the possibility that warmer spring temperatures could dry up critical breeding habitat for waterfowl in the Prairie Pothole Region, a stretch of land between northern Iowa and central Alberta.

Affected wildlife populations can sometimes move into new spaces and continue to thrive. But concurrent human population growth means that many land areas that might be suitable for such "refugee wildlife" are already cluttered with residential and industrial development. The Pew Center on Global Climate Change suggests creating "transitional habitats" or "corridors" that help migrating species by linking natural areas that are otherwise separated by human settlement.

Beyond habitat displacement, many scientists agree that global warming is causing a shift in the timing of various natural cyclical events in the lives of animals. Many birds have altered the timing of long-held migratory and reproductive routines to better sync up with a warming climate. And some hibernating animals are ending their slumbers earlier each year, perhaps due to warmer spring temperatures. To make matters worse, recent research contradicts the long-held hypothesis that different species coexisting in a particular ecosystem respond to global warming as a single entity. Instead, different species sharing like habitat are responding in dissimilar ways, tearing apart ecological communities that have been millennia in the making.

And as wildlife species go their separate ways, humans can also feel the impact. The World Wildlife Fund says that a northern exodus from the United States to Canada by some types of warblers led to a spread of mountain pine beetles that destroy economically productive balsam fir trees. Similarly, a northward migration of caterpillars in the Netherlands has eroded some forests there.

According to Defenders of Wildlife, some of the wildlife species hardest hit so far by global warming include caribou, arctic foxes, toads, polar bears, penguins, gray wolves, tree swallows, painted turtles, and salmon. The group fears that unless we take decisive steps to reverse global warming, more and more species will join the list of wildlife populations pushed to the brink of extinction by a changing climate.

CONTACTS: Pew Center on Global Climate Change, www.pewclimate.org; Defenders of Wildlife, www.defenders.org.

I know that global warming causes extreme weather and melts glaciers and causes sea level to rise, but how does it increase the spread of disease?

Curran Clark, Seattle, WA

Climate change accelerates the spread of disease primarily because warmer global temperatures enlarge the geographic range in which disease-carrying animals, insects, and microorganisms—as well as the germs and viruses they carry—can survive. Analysts believe that, as a result of rising global temperatures, diseases that were limited to tropical areas may increasingly show up in other, normally cooler, areas.

For example, mosquitoes carrying dengue fever traditionally live at elevations no higher than 3,300 feet, but because of warmer temperatures they have recently been seen at 7,200 feet in Colombia's Andes Mountains. And biologists have found malaria-carrying mosquitoes at higher-than-usual elevations in Indonesia in just the last few years. Similarly, mosquitoes carrying avian flu are moving up mountains in Hawaii, threatening remnant native bird populations. These changes happen not because of a few days of extreme heat, but with miniscule increases in average temperature.

Extreme heat can also be a factor, and the nexus of global warming and disease really hit home for North Americans in the summer of 1999, when sixty-two cases of West Nile virus were reported in and around New York City. Dr. Dickson Despommier, a Columbia University public health professor, reports that West Nile virus is spread by one species of mosquito that prefers to prey on birds but will resort to biting humans when its normal avian targets have fled urban areas during heat waves.

"By reproductive imperative, the mosquitoes are forced to feed on humans, and that's what triggered the 1999 epidemic," Despommier says. "Higher temperatures also trigger increased mosquito biting frequency. The first big rains after the drought created new breeding sites." He adds that a similar pattern has been recognized in other recent West Nile outbreaks in Israel, South Africa, and Romania.

Bird flu is likely to spread more quickly as the earth warms up, but for a different reason: a UN study found that global warming—in concert with excessive development—is contributing to an increased loss of wetlands around the world. This trend is already forcing disease-carrying migrating birds, who ordinarily seek out wetlands as stopping points, to instead land on animal farms where they mingle with domestic poultry, risking the spread of the disease via animal-to-human and human-to-human contact.

A congressional report predicts that global warming will cause or increase incidences of malaria, dengue fever, yellow fever, encephalitis, and respiratory diseases throughout the world in coming decades. The assessment also concluded that insect- and rodent-borne diseases would become more prevalent throughout the United States and Europe.

The news isn't good for less-developed parts of the world either. Researchers have found that more than two-thirds of waterborne disease outbreaks (such as cholera) follow major precipitation events, which are already increasing due to global warming.

CONTACT: Natural Resources Defense Council, "Consequences of Global Warming," www.nrdc.org/globalwarming/fcons.asp.

Why does air quality get so bad during heat waves?

Chad Muller, Wellesley, MA

During a heat wave, the combination of heat and sunlight essentially cooks the air, along with all the chemical compounds lingering in it. This chemical soup combines with the naturally occurring nitrogen oxide in the air, creating a smog of ground-level ozone gas. This makes breathing difficult for those who already have respiratory ailments or heart problems and can also make healthy people more susceptible to respiratory infections.

The EPA says urban areas are the most susceptible because of the pollution being emitted from cars, trucks, and buses. The burning of fossil fuels at power plants also emits a considerable amount of smog-making pollution. Geography is also a factor. Broad industrialized valleys penned in by mountain ranges, such as the Los Angeles basin, tend to trap smog, making life miserable for people working or playing outside on hot summer days.

Clean Air Watch reports that intense heat waves in 2006 caused a blanket of smog stretching from coast to coast. Some thirty-eight U.S. states reported an increase in unhealthy air days in July that year. And in some particularly at-risk locales, airborne smog levels exceeded the acceptable healthy standard by as much as a thousandfold.

During heat waves, city dwellers and suburbanites can help reduce smog by using public transit and carpooling to reduce vehicle trips; refueling cars at night to prevent escaping gas vapors from getting cooked into smog by sunlight; avoiding gas-powered lawn equipment; and setting air-conditioning thermostats a few degrees higher to help reduce the fossil fuel burning needed to power them.

The EPA points out that the regulations on power plants and car fuels that have been instituted over the last twenty-five years have significantly reduced smog in American cities. EPA spokesman John Millett says that "ozone pollution concentrations have declined about 20 percent since 1980."

"Long-term we have made improvements . . . but [the 2006] heat wave and the accompanying smog is a very graphic reminder that we still have a significant problem," says Frank O'Donnell of Clean Air Watch.

People should avoid strenuous outdoor activity during heat waves in areas plagued by smog. For more information, check out the government's *Ozone and Your Health* report on the website airnow.gov.

CONTACTS: AirNow, *Ozone and Your Health*, airnow.gov/index.cfm?action=static.brochure; Clean Air Watch, www.cleanairwatch.org.

Are recent cyclones and droughts more evidence of global warming?

Billy Hulkower, Los Angeles, CA

It's a scientific stretch to blame individual storms or droughts on climate change, but many scientists believe that human-induced global warming is increasing the severity and frequency of such weather anomalies. In 2007, the Intergovernmental Panel on Climate Change released a long-awaited report confirming "with 90 percent certainty" that increases in man-made greenhouse gases since the mid-twentieth century are raising the planet's temperature and destabilizing the climate.

Besides hurricanes like Katrina, which have affected the Northern Hemisphere, a number of high-impact tropical cyclones and typhoons have occurred around the world in just the last few years, with Australia's mammoth Cyclone Larry topping the list in terms of intensity. That March 2006 storm battered the North Queensland coast with 180-mile-per-hour winds, causing hundreds of millions of dollars in property damage and virtually wiping out Australia's banana crop. Thanks to Australia's top-notch weather forecasting and emergency preparedness, however, Larry claimed no human lives.

Meanwhile, higher global temperatures have at least worsened if not outright caused drought conditions around the world. The World Wildlife Fund concluded that global warming was a key factor in the severity of Australia's 2002 drought, one of four especially harsh droughts in just the last fifty years. The 2002 drought, which many scientists consider to be still in effect, was particularly memorable as Australians endured higher daytime temperatures than had ever been recorded during any March–November winter season. Besides causing countless bush fires in the Australian Outback, the drought has led to a significant drop in agricultural production, causing hundreds of millions of dollars in economic losses, according to government data.

Australia's conservative government led by Prime Minister John Howard refused to ratify the Kyoto treaty, but Howard was ousted by a Labor candidate, Kevin Rudd, in 2007, and with former Midnight Oil lead singer Peter Garrett in the environmental minister's seat, ratifying climate agreements is a top order of business.

CONTACTS: Asia-Pacific Partnership on Clean Development and Climate, www.asiapacificpartnership.org; Intergovernmental Panel on Climate Change, www.ipcc.ch; World Wildlife Fund-Australia drought report, www.wwf.org.au/publications/drought_report.

What have been the most significant environmental impacts of Hurricane Katrina in New Orleans?

Samantha Gray, Tacoma, WA

Perhaps the longest-lasting impact of Hurricane Katrina has been environmental damage in terms of public health. Significant amounts of industrial waste and raw sewage spilled directly into New Orleans neighborhoods. And oil spills from offshore rigs, coastal refineries, and even corner gas stations have also made their way into residential areas and business districts throughout the region.

An estimated seven million gallons of oil spilled throughout the region. The Coast Guard says that much of the spilled oil has been cleaned up or "naturally dispersed," but environmentalists fear that the initial contamination could affect biodiversity and ecological health for many years to come, further devastating the region's already ailing fisheries, once the economic lifeblood of the area.

Meanwhile, flooding at five Superfund sites (heavily polluted industrial sites slated for federal cleanup) and wholesale destruction along the already infamous Cancer Alley industrial corridor between

New Orleans and Baton Rouge have only served to complicate matters for cleanup officials. The EPA considers Katrina the biggest disaster it has ever had to handle.

Household hazardous wastes, pesticides, heavy metals, and other toxic chemicals also created a witches' brew of floodwater that quickly seeped down into and contaminated groundwater across hundreds of miles. "The range of toxic chemicals that may have been released is extensive," says Johns Hopkins University environmental health sciences professor Lynn Goldman. "We're talking about metals, persistent chemicals, solvents, materials that have numerous potential health impacts over the long term."

Hugh Kaufman, an EPA senior policy analyst, says that environmental regulations in place to prevent the types of discharges that occurred during Katrina were not enforced, making what would have been a bad situation much worse. Unchecked development throughout ecologically sensitive parts of the region put further stress on the environment's ability to absorb and disperse noxious chemicals. "Folks down there were living on borrowed time and, unfortunately, time ran out with Katrina," Kaufman concludes.

To date, recovery efforts have focused on plugging leaks in levees, clearing debris, and repairing water and sewer systems. Officials cannot say when they will be able to concentrate on longer-term issues such as treating contaminated soil and groundwater.

Meanwhile, financially strapped state and local agencies are slowly removing contaminated buildings, many of which harbor mold and viruses that can still make people sick. Sadly, the Federal Emergency Management Agency said in early 2008 that it would have to move Gulf Coast hurricane victims out of the thirty-five thousand trailers it had provided to them because of high levels of formaldehyde.

CONTACTS: EPA's Response to 2005 Hurricanes Website, www.epa.gov/katrina; *E* magazine, "The Toxic Legacy of Hurricane Katrina," www.emagazine.com/?issue=125&toc.

What is an "urban heat island," and does it have anything to do with global warming?

Max, via e-mail

Why are cities warmer than the surrounding countryside? Welcome to "urban heat islands." Unlike global warming, which causes a worldwide rise in temperatures, heat islands occur at the local level. According to the EPA, many cities and suburbs have air temperatures up to 10 degrees Fahrenheit warmer than their neighboring areas.

Heat islands form as cities pave over the natural world, replacing it with buildings and sidewalks. These changes contribute to higher urban temperatures in a number of ways. For one, displacing trees and removing soil and vegetation takes away the natural cooling effects that shading and water evaporation from soil and leaves ordinarily provide. Meanwhile, tall buildings and narrow streets can heat the air trapped between them and reduce airflow. And waste heat from vehicles, factories, and air conditioners adds warmth to the surroundings, further exacerbating the heat-island effect.

The intensity of a heat island will also depend upon its topography, its proximity to water bodies, and local weather and climate. Urban heat islands can also impact local weather, altering local wind patterns, spurring the development of clouds and fog, increasing the number of lightning strikes, and influencing rates of precipitation.

Urban heat islands are not climate change, but during the summer months they can contribute to global warming. The increased use of air-conditioning and refrigeration needed to cool indoor spaces in a heat-island city, for example, results in the release of more of the heat-trapping greenhouse gases that cause global warming. Furthermore, the poor air quality that results from this increased energy usage can affect our health, aggravating asthma and promoting other respiratory illnesses.

Costs are impacted too. The Heat Island Group, a research and advocacy organization that works to educate the public and policy makers about the heat-island effect, estimates that the city of Los Angeles spends about $100 million per year in extra energy costs to offset its heat-island effect.

The heat-island effect can be reduced through the use of white and light-colored construction materials (including white roofing materials) in buildings, which will work to reflect the sun's heat skyward rather than absorb it, as dark surfaces tend to do. Also, preserving or creating pockets of green space and vegetation will help to cool areas naturally. A national program called Cool Communities, coordinated by American Forests and supported by the Department of Energy, works on these issues. Another useful practice is the creation of "green roofs" or rooftop gardens, in which roofs are partially or completely covered with vegetation and soil, or a growing medium, planted over a waterproofing layer.

CONTACTS: American Forests, www.americanforests.org; EPA heat-island effect information, www.epa.gov/heatislands; Heat Island Group, http://eetd.lbl.gov/heatisland.

I recently heard the term "carbon sequestration" in relation to climate change. What is it and how can it help stave off global warming?

Bob Whelan, Pawtucket, RI

The most common natural example of carbon sequestration, which is the taking in and storing of carbon dioxide, is the photosynthesis process of trees and plants. Because their growth soaks up the carbon dioxide that would otherwise rise up and trap heat in the atmosphere, trees and plants are important players in efforts to stave off global warming.

Environmentalists cite this natural form of sequestration as a key reason to preserve the world's forests and other undeveloped lands where vegetation is abundant. And forests don't just absorb and store large quantities of carbon dioxide; they also produce large quantities of oxygen as a by-product, leading people to refer to them as the "lungs of the earth."

The Western Canada Wilderness Committee says the billions of trees in the boreal forest of the Northern Hemisphere that stretches from Russian Siberia across Canada and into Scandinavia absorb vast amounts of carbon dioxide as they grow. Likewise, the world's tropical forests play an important role in naturally sequestering carbon dioxide. Environmentalists see preserving and adding to the world's forest canopy as the best natural means for minimizing the impact of global warming caused by the 5.5 billion tons of carbon dioxide generated by factories and automobiles each year.

Engineers are developing man-made ways to capture the carbon dioxide spewing from coal-fired power plants and industrial smokestacks and sequester it by burying it deep within the earth or the oceans. The Bush administration, for one, embraced carbon sequestration as a means of mitigating U.S. CO_2 emissions, spending forty-nine million dollars annually on research and development. The United States also began funding research on approaches to stemming Chinese emissions, which recently overtook the American sizable contribution.

CONTACT: Intergovernmental Panel on Climate Change, *Special Report on Carbon Dioxide Capture and Storage,* www.ipcc.ch/ipccreports/srccs.htm.

Which trees are best to plant to combat global warming?

Tim C., Perrineville, NJ

Trees in general absorb and store CO_2 before it has a chance to reach the upper atmosphere, where it can help trap heat around the earth's surface. While all living plant matter absorbs CO_2 as part of photosynthesis, trees process significantly more than smaller plants due to their large size and extensive root structures. In essence, trees, as kings of the plant world, have much more woody biomass to store CO_2 than smaller plants do, and as a result are considered nature's most efficient "carbon sinks."

Tree species that grow quickly and live long are ideal carbon sinks. Unfortunately, these two attributes are usually mutually exclusive. Foresters interested in maximizing the absorption and storage of CO_2 (known as "carbon sequestration") usually favor younger trees that grow quickly. However, slower growing trees can store much more carbon dioxide over their significantly longer lives.

Scientists are busy studying the carbon sequestration potential of different types of trees in various parts of the United States, including eucalyptus in Hawaii, loblolly pine in the Southeast, bottomland hardwoods in Mississippi, and poplars in the Great Lakes. "There are literally dozens of tree species that could be planted depending upon location, climate, and soils," says Stan Wullschleger, a researcher at Tennessee's Oak Ridge National Laboratory.

The U.S. Forest Service has studied the use of trees for carbon sequestration in urban settings across the country and lists the common horse chestnut, black walnut, American sweet gum, ponderosa pine, red pine, white pine, London plane, Hispaniolan pine, Douglas fir, scarlet oak, red oak, Virginia live oak, and bald cypress as examples of trees especially good at absorbing and storing CO_2. The Forest Service advises urban land managers to avoid trees that require a lot of maintenance, as the burning of fossil fuels to power equipment like trucks and chainsaws will only erase the carbon-absorption gains otherwise made.

Ultimately, trees of any shape, size, or genetic origin help absorb CO_2. Most scientists agree that the least expensive and perhaps easiest way for people to offset the CO_2 they generate in their everyday lives is to plant a tree, any tree, as long as it is appropriate for the given region and climate. To help with larger tree-planting efforts, you can donate money or time to the Arbor Day Foundation or American Forests in the United States, or to Tree Canada in Canada.

CONTACTS: American Forests, www.americanforests.org; Arbor Day Foundation, www.arborday.org; Tree Canada, www.treecanada.ca.

What are the implications of the increased breakup of Antarctica's large floating ice shelves in recent years?

Gaertner Olivier, Brussels, Belgium

Ice shelves are thick plates of ice that float on the ocean around much of Antarctica. Snow, glaciers, and ice floes feed these large plates in the colder months. In warmer periods, surface melting creates standing water that leaks into cracks and speeds the breaking off (calving) of icebergs, decreasing the continent's mass in a natural cycle as old as Antarctica itself.

"Large icebergs calve off on a fairly regular basis from the larger ice shelves in Antarctica," says Dr. Ted Scambos, a research associate at National Snow and Ice Data Center. "This is a part of their normal evolution."

The only effect of such calving that scientists are sure about is that they are changing the outline of Antarctica. The breakup of the ice shelves, which account for about 2 percent of the continent's landmass, does not have any measurable effect on sea levels. "Since an iceberg floats in ocean water, and much of it is below the surface, it is already displacing the same volume of water it will contribute when it eventually melts," Scambos explains.

But while such calving activity may not be new, it has increased over the last thirty years, with larger and larger chunks breaking off from Antarctica, after which they float free in the ocean and break up into successively smaller pieces. One especially large iceberg, a chunk the size and shape of New York's Long Island and dubbed B15A by researchers, broke off from Antarctica's Ross Ice Shelf in 2000 and later collided with the continent's Drygalski Ice Tongue (a long shelf of ice extending out to sea from the mainland). The iceberg itself remained intact, but a city-sized chunk of the ice tongue broke off and floated free.

Most researchers suspect that recent increases in calving are linked to warming surface air temperatures as a result of human-induced climate change. British glaciologist David Vaughan says, "There's no doubt that the climate on the Antarctic Peninsula has warmed significantly over the last few decades. What we're seeing now are changes only just working through to glaciers and ice sheets." Scambos says that, as Antarctic summer temperatures continue to increase, the process can be expected to become more widespread and could begin to significantly increase sea levels around the world.

Even a relatively small rise in sea level would make some densely settled coastal areas uninhabitable. The Intergovernmental Panel on Climate Change, an international group of climatologists, predicts a global sea-level rise of less than three feet by 2100 but also warns that global warming during that time may lead to irreversible changes in the earth's glacial system and ultimately melt enough ice to raise sea levels many more feet in coming centuries. Some two hundred million people inhabit low-lying areas in countries like Vietnam, Bangladesh, China, India, and the Philippines and could be displaced, leading to a major international refugee crisis.

CONTACTS: Intergovernmental Panel on Climate Change, www.ipcc.ch; NASA's iceberg collision page, www.nasa.gov/vision/earth/lookingatearth/ice_berg_ram.html.

How are coral reefs faring around the world?

Debby Greco, Canton, CT

Not so well, unfortunately. The World Resources Institute reports that coral reefs around the world are dying or disappearing at an alarming rate.

Lining sixty thousand miles of shoreline along 109 countries, reefs and their related fisheries, marshlands, and lagoons are home to more than a quarter of all fish species on Earth. An estimated 25 percent of coral reefs have already disappeared and an estimated 67 percent of all remaining coral reefs are endangered today. In Southeast Asia, 88 percent of the reefs are at risk. In the Florida Keys, more than 90 percent of the reefs have lost their living coral cover since 1975.

The Planetary Coral Reef Foundation, which monitors the health of coral reefs worldwide, says the greatest threat to coral reefs thus far has been coastal development resulting from human population expansion. Over the last thirty years, this trend has profoundly increased the amount of freshwater runoff into coastal areas. Known collectively as "nonpoint source pollution," this runoff has carried large amounts of sediment, sewage, and chemicals from land-clearing areas, agricultural areas, and septic systems into the reefs. The resulting water pollution decreases the amount of light reaching the corals, choking the life out of these fragile structures.

Increases in both commercial and sport fishing, enhanced by ever-improving technologies, have also taken a toll on reef health by removing many of the large fish that, when healthy and plentiful, keep fragile reef ecosystems in balance.

Scientists studying coral health are most concerned about the impact of a somewhat newer threat: bleaching as a result of climate change. Indeed, global warming is changing the surface temperatures of ocean waters faster than corals can adapt. "Coral reefs are so sensitive to temperature change that it seems inevitable that many will die as a result of global warming as well as all the other terrible things that are happening to them," says Rod Fujita, a marine biologist with the Environmental Defense Fund. Furthermore, coral reefs' very sensitivity to environmental changes makes them an early warning system with regard to the overall declining health of the world's oceans.

The Coral Reef Alliance is working toward the establishment of a comprehensive global map of living coral reefs to serve as a baseline for learning how fast we are losing them and how we can stem the decline. Also, a new NASA program is monitoring coral reefs by satellite to try to put some of the puzzle pieces together.

CONTACTS: Coral Reef Alliance, www.coralreefalliance.org; Environmental Defense Fund, www.environmentaldefense.org; NASA, www.nasa.gov; Planetary Coral Reef Foundation, www.pcrf.org; World Resources Institute, www.wri.org.

Someone told me that methane gas emitted by cows is a major contributor to global warming. I thought it was a joke, but is this true?

David Rietz, Goose Creek, SC

It's all too true! Cows produce copious amounts of methane gas, and methane accumulation in the earth's atmosphere has nearly doubled around the globe over the past two hundred years. Scientists believe that rising concentrations of methane, which absorbs and sends infrared radiation to the earth, are causing changes in the climate and contributing to global warming.

Livestock animals naturally produce methane as part of their digestive process, belching it while chewing cud and excreting it in their waste. According to the Worldwatch Institute, about 15 to 20 percent of global methane emissions come from livestock. John Robbins, author of *The Food Revolution* and *Diet for a New America*, says that methane is twenty-four times more potent a greenhouse gas than CO_2.

The Department of Agriculture says that animals in the U.S. meat industry produce sixty-one million tons of waste each year, which is 130 times the volume of human waste produced, or five tons for every U.S. citizen. In addition to its impact on climate, hog, chicken, and cow waste has polluted some thirty-five thousand miles of rivers in twenty-two states and contaminated groundwater in seventeen states.

In 2006, the UN released a report entitled *Livestock's Long Shadow* that showed meat production accounting for some 18 percent of all greenhouse gas emissions—incredibly, more than the entire transportation sector.

It takes seven pounds of corn to add a pound of weight to a cow. That's why two hundred million acres of land in the United States is devoted to raising grains, oilseeds, pasture, and hay for livestock. Raising livestock consumes 90 percent of the U.S. soy crop, 80 percent of its corn, and 70 percent of its grain. David Pimentel of Cornell points out: "If all the grain currently fed to livestock in the U.S. was consumed directly by people, the number who could be fed is nearly 800 million."

Ronnie Cummins, national director of the Organic Consumers Association, says that a food chain with meat at its top is unsustainable not only as a major contributor of greenhouse gases but also with regard to inefficient dedication of large amounts of acreage to livestock grazing. The USDA, for example, says that growing the crops necessary to feed farmed animals requires nearly 80 percent of America's agricultural land and half of its water supply.

Unfortunately, environmental problems associated with livestock rearing are not limited to the United States. According to the international environmental journal *Earth Times*, meat production grew more than fivefold worldwide during the latter half of the twentieth century. And as intensive "factory" farming methods of raising livestock spread from the United States to other countries—many with regulatory monitoring and enforcement standards far worse than our own—

this form of pollution is sure to play an increasingly larger role in environmental problems.

CONTACTS: Organic Consumers Association, www.purefood.org; Worldwatch Institute, www.worldwatch.org.

Why are beaches and coastlines eroding and what can be done about it?

Jesus Lopez, Santa Maria, CA

Beach erosion has both human and natural causes. The process of erosion carries beaches out to sea, but it also created them over millions of years from the rock-strewn shores that originally covered our planet.

"Without erosion, we would not have the beaches, dunes, and highly productive bays and estuaries that owe their very existence to the presence of barrier beaches," says Jim O'Donnell, a coastal processes specialist with the Sea Grant Program at Woods Hole Oceanographic Institution.

Sand moves naturally through the actions of wind and the tide, but it is helped along by human actions, and the beach erosion problem is growing dramatically. The main causes are overbuilding right to the water's edge (a practice protected by federal flood insurance), rapid rises in sea level exacerbated by global warming, a gradual sinking of coastal land, and inept attempts to fix the problem.

Scott L. Douglass, author of *Saving America's Beaches* and a professor at the University of South Alabama, worked his way through college lifeguarding on the New Jersey shore. Like many beach experts, he's a major critic of the erosion-promoting effects of jetties, seawalls, and dredging. Human activity has removed "more than a billion cubic yards of sand from the beaches of America, enough to fill a football field over 100 miles high," he points out. Douglass prefers beach replenishment, which he says "adds sand to the system," but he acknowledges that, with sea levels rising at a rate of six inches every hundred years, beaches may not be able to keep up.

Rising sea level means that wetlands and other low-lying lands get inundated, beaches erode, flooding intensifies, and the salinity of rivers, bays, and groundwater tables increases. Sea level is rising more rapidly along the U.S. coast than worldwide, according to the EPA. In the next century, a two-foot rise is likely, but a four-foot rise is possible; and sea level will probably continue to rise for several centuries, even if global temperatures were to stop rising.

Orrin Pilkey, who directs the Program for the Study of Developed Shorelines at Duke University, believes that in many cases it would actually be cheaper to move buildings back from the water's edge than to fund ten to twenty years of constant beach replenishment, but his ideas have not had many takers among shoreline communities. Some states and localities in the United States and around the world have setback requirements restricting development on the shoreline. Protecting and restoring natural barriers to erosion like dunes, wetlands, and vegetation close to shore are also natural, low-cost ways to fight erosion.

CONTACTS: EPA, "Coastal Watershed Factsheets," www.epa.gov/owow/oceans/factsheets; Western Carolina University, Program for the Study of Developed Shorelines, (828) 227-3822, http://psds.wcu.edu; Woods Hole Oceanographic Institution, (508) 289-2252, www.whoi.edu.

9

OPEN ROAD

Transportation, Travel, and Cutting Our Carbon Load

Have you ever seen a car commercial set in actual traffic conditions? Instead, the gleaming vehicle swoops down a deserted country road as its young driver enjoys the perfect bonding experience. That's our love affair with the automobile personified. In real life, love dies a little every time we sit breathing smog in bumper-to-bumper traffic. Throw in the dual whammies of global warming and high prices at the pump, and it's not surprising that Americans are turning to smaller, more fuel-efficient cars.

We're finally coming around to hybrid cars and biodiesel, due in equal parts to concern for our diminishing bank accounts and for our diminishing resources. And maybe we're finally starting to wake up from our commuting nightmares and to recognize the freedom offered by public transportation and even occasional bicycle trips. Traveling with an eco-conscience doesn't mean we'll never hop aboard another fuel-burning jet to reach an exotic locale, but it may change where we land and how we act when we get there. Simply by reusing towels and supporting local farmers and crafters, we can begin to make a difference with how we spend our dollars.

What are the most environmentally friendly and high-mileage cars on the market today? Also, are the batteries in hybrid cars recyclable?

Shiela Gosselin, via e-mail

According to the annual environmental rating of the best and worst cars by the American Council for an Energy Efficient Economy

(ACEEE), Honda, Nissan, and Toyota models lead the pack as the world's "greenest" automobiles. Not surprisingly, top honors go to a best-selling hybrid gasoline-electric vehicle, the Toyota Prius, which pairs an efficient electric motor with a gasoline engine to save gas and minimize emissions.

Even greener, actually, is the Honda Civic GX natural gas car, but that's being sold to a smaller niche market. Ford has a pair of "greener" choices too: the Focus sedan and Escape hybrid. Making its first appearance on the 2008 list is the tiny smart fortwo two-seater, which gets thirty-three miles per hour in city driving and forty-one on the highway.

To determine a car's rankings, ACEEE factors in fuel efficiency plus the pollution generated by a given vehicle based on EPA emissions ratings. Other "greener" choices include the small Honda Fit, the sporty Mazda MX-5 Miata, and the Hyundai Sonata.

Hybrid advocates insist that the nickel–metal hydride batteries found in the Toyota Prius, Ford Escape, and other hybrids contain far fewer pollutants than the lead-acid types present in traditional gas-powered cars. Furthermore, carmakers are keen to keep such batteries out of landfills, with Toyota even offering to buy back spent hybrid batteries for two hundred dollars so it can recycle them.

According to Toyota: "Every part of the battery, from the precious metals to the plastic, plates, steel case, and the wiring, is recycled." Bradley Berman of hybridcars.com reports, "Honda collects the battery and transfers it to a preferred recycler to follow the prescribed process: disassembling and sorting the materials; shredding the plastics; recovering and processing the metal; and neutralizing the alkaline material before sending it to a landfill." Automakers are scrambling to create smaller, more efficient and less-toxic batteries for hybrids and other vehicles, Berman reports.

Another option for green consumers is a diesel car that runs on biodiesel, a fuel derived from renewable crops (and which works seamlessly in most diesel engines). A recent *AutoWeek* report concluded that a biodiesel-powered Volkswagen Jetta TDi has the best overall fuel economy of any new car on the road under "real-world driving conditions" (including, among other things, traffic congestion, use of air-conditioning, and high speeds). In *AutoWeek*'s test-drive comparison, the Jetta TDi achieved nearly fifty miles per gallon using B20 biodiesel (two parts vegetable oil, eight parts regular diesel), edging out even Toyota's Prius, which scored forty-two miles per gallon using gasoline.

The EPA revised its mileage testing procedures for the 2007 model year, getting them more in line with real-world driving conditions. As a result, fuel economies displayed on window stickers are moving downward. Some cars, especially smaller vehicles and hybrids, will lose as much as 12 percent in their ratings.

CONTACTS: American Council for an Engery Efficient Economy's *Green Book* online, www.greenercars.org; hybridcars.com, www.hybridcars.com.

What are the best eco-friendly vehicle choices for those of us who need a pickup or sport-utility vehicle (SUV)? We are about to replace two older trucks with one that is more fuel efficient.

Barbara Roemer, via e-mail

Fuel efficiency has not typically been the calling card of pickup trucks and sport-utility vehicles (SUVs). Small hybrid gasoline-electrics are all the rage now among commuters looking to save money at the pump, but similar technology has been slower to gain traction in the "light truck" category. Carmakers have made strides in recent years, though, to meet the growing demand for vehicles of all kinds that will sip and not gulp.

Hybrid versions of General Motors' Chevy Silverado and GMC

Sierra 4x4s have been available since 2005 and get about 18 miles per gallon (mpg) in the city and 21 on the highway. There's no huge advantage; the slightly cheaper nonhybrid versions get 15/19 mpg. GM claims that those paying the hybrid premium will get back that extra investment in fuel savings over three to five years.

Toyota reportedly has plans for hybridizing its full-size pickup line. The company recently unveiled its FTX concept truck, a large 4x4 hybrid pickup, hinting that technology developed for the project will likely end up in its current full-size Tundra pickup. Another likely hybrid is Honda's popular (and very big) Ridgeline pickup.

Ford currently leads the charge for fuel-efficient SUVs with its Escape hybrid model, which gets 34/30 mpg. Ford makes similar SUV hybrids under its Mercury and Mazda brands. Toyota's midsize Highlander hybrid SUV clocks in at 27/25 mpg, and the larger and more luxurious Lexus RX-400 Hybrid gets 27/24 mpg. All these vehicles post significantly better fuel-efficiency ratings than their nonhybrid counterparts but also cost more up front.

Replacing an older truck with a newer model, especially a hybrid, will almost always guarantee better fuel economy, but it might not be the most environmentally sensitive way to go. Some experts would argue for keeping the old truck and fixing and tuning it up, thus preventing another new vehicle from hitting the roads while an old one clogs up the junkyard. Repairing an old vehicle is usually cheaper than buying a new one, though it is difficult to quantify the cost of ongoing maintenance hassles.

CONTACT: EPA fuel economy information, www.epa.gov/fuel economy.

I'm familiar with the hybrid cars now widely available, but what ever happened to the purely electric cars that were around ten years ago?

Peter Zilly, Bellingham, WA

The main problem with electric vehicles (EVs) is that, to date, they've been able to only go so far on battery power. Charges on most last just fifty miles or so, so you're in trouble if you need to go farther or run out of juice somewhere between electric outlets. Hybrids, with side-by-side electric and gas motors, never need to be plugged in and instead use the motions of their gas-powered engines (as well as those of the car's wheels and brakes) to keep their batteries charged at all times. And with the huge existing infrastructure of gas stations, refueling is always as easy as pulling over to fill up.

EV advocates tout their alternative vehicles as short-distance commuter cars. At a fifty-mile range, most electric cars could make such short trips without the need for recharging. One need only plug their vehicle into an electric outlet in the garage overnight to charge up the battery for the morning commute and, if needed, then plug it in at the office for the return trip later.

But most people want more from their cars than just the daily commute—and gassing up takes minutes whereas recharging takes hours—so sufficient demand hasn't materialized. Hybrids are as versatile as conventional cars, and the coming "plug-in hybrids" promise to substantially increase efficiency, to perhaps one hundred miles per gallon or more, by using the electric motor exclusively for short runs and commutes and the gas engine only for long trips.

Even though EVs are not currently in vogue, engineers are busy working to improve them. Technological advances in battery life and engine efficiency mean that electric vehicles may be able to roam farther than ever before. According to the website EV World, drivers looking to go electric will soon have a few options.

In 2008, California-based Tesla Motors began production of the Tesla Roadster, which can go two hundred or more miles on a charge. The hundred-thousand-dollar car, which aims to go from zero to sixty in four seconds, also offers "regenerative braking" technology to capture energy, as also seen on the hybrids.

Commuter Cars Corporation in Spokane, Washington, offers the Tango in three models: Tango 600 (a kit you have to assemble) and the Tango 100 and 200 models (fully assembled). Actor George Clooney was Commuter Cars' first customer. The Tango can only go sixty to eighty miles on a charge but boasts of an ability to go from zero to sixty in four seconds and attain a top speed of 150 miles per hour.

California-based AC Propulsion is working with Toyota on a Scion electric conversion, and Cleanova, based in France, is developing an electric Renault Kangoo, a popular European car.

There's another approach to the electric-battery car, a "series" hybrid that uses a small gasoline engine solely to keep the batteries charged. General Motors is working on a sedan using that technology, the Chevrolet Volt, and intends to have it on the market soon. Pete Savagian, engineering director of GM Hybrid Powertrain, says, "Reducing our dependence on petroleum requires vehicles that provide the petroleum-free benefits we know electric vehicle drivers are passionate about, but we also need to offer the flexibility to be able to drive hundreds of miles at a time between fill-ups. We believe that an extended-range electric like the Volt is that kind of vehicle."

The practicality of EVs could hinge on whether your utility runs a dirty coal-fired plant. If so, tapping that power could mean creating more pollution than driving a gasoline-powered car. But progress in renewable energies may well solve that problem and help usher in a new era for electric vehicles.

CONTACTS: Commuter Cars Corporation, www.commutercars.com; EV World, www.evworld.com; Tesla Motors, www.teslamotors.com.

As I understand it, hybrid cars make use of an electric motor that never needs to be plugged in. But what's up with the proposed "plug-in" hybrids I've been hearing about?

Jen Seminara, Omaha, NE

The mass-market gasoline-electric hybrids made by Toyota, Honda, and others make use of an electric engine right under the hood next to the gas engine. That electric motor creates fuel economy by kicking into use during idling, backing up, slow traffic, and to maintain speed after the gas engine has been employed for acceleration. The car doesn't need to be plugged in because the on-board electric battery is constantly being charged by the gas engine and by the motion of the wheels and the brakes.

The "plug-in" hybrids take this technology a step further. By adding the ability to charge up from a standard household outlet, typically overnight, they relegate the gas engine to backup status and instead let the electric motor do most of the work.

Such "gas-optional" cars (if you don't take long trips you can rely entirely on the electric motor) can be twice as fuel efficient as hybrids, which already get double the gas mileage of traditional vehicles. Additionally, they say, powering plug-in hybrids with wall sockets results in far less pollution (from the power plants providing the electricity) than an equivalent gasoline-powered car spews out its tailpipe. And plug-in hybrids recharged from rooftop solar power systems might come close to being the world's first mass-market "zero emission" vehicles, requiring no power from the grid at all.

Convincing a skeptical American public that plug-in hybrids are the way of the future is the challenge of a loose network of advocacy groups led by the California Cars Initiative (CalCars).

"It's like having a second small fuel tank that you always use first—only you fill this tank at home with electricity at an equivalent cost of under $1 per gallon," reports the CalCars website. The organization goes on to explain that with gas prices at more than $3 per gallon, traditional cars cost 8 to 20 cents per mile, but plug-in hybrids used for all-electric local travel and commuting would cost only 2 to 4 cents per mile.

To convince the carmakers to build plug-ins, CalCars has built showcase examples that achieve one hundred miles per gallon (based on Toyota's Prius). But both General Motors and Toyota have gotten the message and are working on production plug-ins. DaimlerChrysler has built a handful of plug-in prototypes based on its fifteen-passenger Mercedes-Benz Sprinter van, but has no production plans. And a growing list of state and local governments say they would seriously consider converting their fleets to plug-in hybrids when such vehicles are available.

CONTACT: California Cars Initative, (650) 520-5555, www.calcars.org.

What exactly are "partial zero emission vehicle" (PZEV) cars? Someone told me they were very clean and on the market now.

Thomas Lyons, Jamaica Plain, MA

Thanks to rigorous auto emissions standards in California—where regulators are trying to clean up the worst air in the country—no less than a dozen car companies now offer partial zero emission vehicles (PZEV) for sale in the United States. These cars run on gasoline and don't necessarily get better mileage than their traditional counterparts, but they do produce much cleaner emissions (as much as 90 percent cleaner than the average car) by controlling exhaust gases with sophisticated engine controls and advanced catalytic converters. In some cases, the exhaust of a PZEV is cleaner than the ambient air!

Most auto pollution is released while a car is warming up and the catalytic converter is still cold. But PZEVs, through the use of lightweight steel and aluminum components, computerized valve timing, and other advanced engineering technologies, heat the catalytic converter quickly, which reduces emissions significantly. These reduced emissions qualify the cars as "low-emission vehicles" (LEVs) in the "clean car states" on the East and West coasts, each of which requires automakers to sell a certain percentage of "green" cars.

Environmentalists are optimistic that the fast-growing fleet of PZEVs on America's roads will have a much larger and more positive impact on environmental quality than the even cleaner running gasoline-electric hybrids, which are still niche vehicles. For every hybrid Prius sold by Toyota since it was introduced in 2000, Ford has sold three PZEV Focuses. Even Subaru has sold more than one hundred thousand of them.

What's perhaps most striking about PZEVs is their "stealth" marketing, especially in light of all the attention being paid to the hybrids and to the coming hydrogen fuel cell cars. All new versions of Ford's popular Focus model, for example, meet PZEV standards, but consumers wouldn't know it unless they were to ask. Compared to a similar traditional car, the PZEV Focus produces 97 percent fewer hydrocarbon and nitrogen oxide emissions and 76 percent less carbon monoxide.

Car buyers wanting a PZEV have to shell out a few hundred dollars extra for the greener technology, but choices include the BMW 325i, Dodge Stratus, Chrysler Sebring, Honda Accord, Hyundai Elantra, Mitsubishi Galant, Nissan Sentra, the Subaru Outback, Legacy, and Forester, Toyota Camry, Volkswagen Jetta, Volvo S60 sedan and V70 wagon, and of course the Ford Focus. Consumers in the "clean car states" should be able to order any of the PZEV models at local auto dealers. Only the Ford Focus and three Subaru models are readily available in all fifty states but, says *Green Car Journal*, "It's just a matter of time until the rest of the country catches up and we can all breathe a bit easier."

CONTACTS: California's DriveClean website, www.driveclean.ca.gov; Ford Focus, www.fordvehicles.com/cars/focus; Green Car Journal, www.greencar.com.

I understand that you can run a diesel car on used cooking oil. Why would I want to do that, and how would I convert such a vehicle to do so?

Benjamin Crouch, Boston, MA

Biodiesel is one of the cleanest-burning alternative fuels available today. Diesel engines that run on biodiesel emit substantially less carbon monoxide and soot compared to engines that run on regular diesel fuel. They also emit almost no sulfur oxide or sulfates, major components of acid rain. Biodiesel also won't add CO_2 to the atmosphere, and as such is not a significant contributor to global warming. According to the National Biodiesel Board, it is the only alternative fuel to meet Clean Air Act testing requirements.

Biodiesel is green friendly because it comes from plants, primarily soybeans and rapeseed (also marketed for cooking as canola oil). Like all plants, these crops absorb CO_2 during their lifetimes. Later, when burned, they emit only about as much CO_2 as they absorbed before they were harvested, thus making them what is called "carbon neutral."

Biodiesel can power any standard diesel automobile without major engine modifications. While it can be used in its pure form, biodiesel can also be mixed with less-expensive standard diesel fuel to significantly reduce pollution compared to using standard diesel alone. Those who want to go a step further can retrofit their existing diesel-powered cars to run on straight vegetable oil, as some environmentally ambitious drivers have done, obtaining oil from the fryer of their local fast-food or Chinese restaurant.

An estimated five thousand North Americans have converted their diesel cars or trucks to run on vegetable oil in the last few years alone. Those who do so usually make a deal with a local eatery willing to hand over its used cooking oil at the close of the business day.

Drivers willing to spend between four hundred and a thousand dollars on a conversion kit from one of two leading vendors, Missouri-based Golden Fuel Systems and Massachusetts-based Greasecar Vegetable Fuel Systems, can make the switch. And fryer-friendly restaurants are just about the only economical fuel source right now. Buying cooking oils at the supermarket would be costly, and consumers shouldn't expect to find filling stations pumping vegetable oil anytime soon.

The conversion kits are only for diesel vehicles, as gasoline engines do not tolerate vegetable oil as a fuel. The conversion involves replacing and moving hoses and leads as well as adding a separate fuel tank for the vegetable oil, so it is best handled by a trained mechanic. Drivers should know that a converted vehicle does need a small amount of regular diesel fuel to get started, because at normal or cold temperatures vegetable oil is too thick to properly ignite. But the vehicle can switch over to vegetable oil once it is warmed up and the heat inside the engine loosens its thickness so it can run through efficiently.

Biodiesel vendors have set up pumping stations across North America, although they tend to be few and far between. Canadians can locate biodiesel stations at the website of the Canadian Renewable Fuels Association; Americans can consult the website of the National Biodiesel Board.

CONTACTS: Canadian Renewable Fuels Association, www.greenfuels.org; Golden Fuel Systems, www.goldenfuelsystems.com; Greasecar Vegetable Fuel Systems, www.greasecar.com; National Biodiesel Board, www.biodiesel.org.

Are there any environmental benefits to diesel-powered cars?

Bill Darcy, Concord, NH

Choosing the high fuel economy of a diesel-powered vehicle, such as the Volkswagen Golf GL TDI, might seem like a no-brainer for anyone looking for a new car that will consume less gas and money. But although the TDI gets far better gas mileage than a similar gasoline-powered Golf, its diesel engine emits considerably more harmful pollutants into the air.

In fact, a Swedish study found that diesel-powered cars in India had twice the cancer potency level of gasoline-powered vehicles in that country. The results were supported by German research. And the Union of Concerned Scientists reports that in California diesel exhaust accounts for 70 percent of the cancer risk from that state's polluted air. According to the EPA, the particulates, or soot, in diesel exhaust cause a host of health problems, including irritation of the eyes, nose, and throat; heartburn; headaches and lightheadedness; and asthma and lung disease.

Fortunately, the diesel picture in the United States is improving with the introduction, beginning in 2006, of ultralow-sulfur diesel fuel. All highway diesel will be low-sulfur by the end of 2010. One new vehicle taking advantage of that mandate is the Mercedes-Benz E320 BLUETEC, a diesel (introduced for the 2007 model year) that environmentalists have to take seriously. The BLUETEC cleans up its act with a chemical reaction caused by injecting urea into exhaust gases before they reach the catalytic converter.

The really zippy acceleration (almost as fast as the gas-powered E350) and spectacular fuel economy (thirty-two miles per gallon on the highway and twenty-three in town) are also part of the charm. Is there a downside? Well, you do have to keep the wiper-fluid-size urea tank topped up, and you can't buy BLUETECs just anywhere. But by the 2010 model year, you should be able to buy an E320 BLUETEC in all fifty states.

CONTACTS: EPA, "Green Vehicle Guide," www.epa.gov/greenvehicles; Union of Concerned Scientists, www.ucsusa.org.

Are there any car-free cities in the world?

Elizabeth Vales, Cleveland, OH

Since the dawn of the automobile age, city dwellers worldwide have been choking on exhaust fumes and tempting fate every time they enter a crosswalk. J. H. Crawford, author of *Carfree Cities*, says as much as 70 percent of downtown space in most American and European urban centers today is dominated by traffic lanes, parking lots and garages, gas stations, drive-through banks and burger stands, and of course car dealerships.

If you live in a sprawling city such as Houston or Atlanta, you probably spend 22 percent of your annual income on automobile and related expenses. Cars aren't so great for business, either: A recent study of thirty-two German cities concluded that fewer cars allowed into a city meant increased foot traffic and more retail sales.

Carfree.com, the online companion to Crawford's book, offers a large listing of car-free places throughout the world, organized into three categories: those completely or predominantly car free; those with large areas that are car free; and those with limited automobile traffic. In the United States, essentially car-free locations (though not cities) include Mackinac Island, a resort island on Lake Huron that uses horses-drawn buggies for its transportation, and Fire Island on Long Island in New York. Fire Island makes use of small boats for short dock-to-dock travel and wagons for wheeling the groceries home. It also has a lengthy network of boardwalks connecting homes on the beach to one another and to the docks.

Most car-free places are in Europe, the largest being Venice, where a canal system takes the place of streets, and movement is on foot or by boat. Giethoorn, in the Netherlands, also relies on canal-boat transportation. Some alpine resorts in Switzerland, such as Zermatt and Braunwald, are car free as well. A unique location is Louvain-la-Neuve, a university town in Belgium where streets for cars lie beneath separate streets for pedestrians. There are also car-free cities in Morocco where, according to carfree.com, they have succeeded in preserving much of the medieval-style narrow streets. They are "for practical reasons, substantially car-free, although not always motorcycle-free," says the website.

There are also car-free cities and areas in much of the developing world, though this is mainly due to poverty. But increasingly, the four billion inhabitants of the developing world seem eager to adopt Western patterns, and automobile use is growing. In India, for example, according to the UN, the number of cars has been doubling every seven years.

CONTACTS: *Carbusters* magazine, www.carbusters.org; Carfree, www.carfree.com; World Carfree Network, www.worldcarfree.net.

Where I live, our highways are "parking lots." Isn't this an ideal situation for public transit? Why isn't it happening?

John Moulton, Stamford, CT

In fact, an increasing number of public transit options are available in North America, but if you're idling alone in bumper-to-bumper traffic you might not know it. Indeed, lack of knowledge about public transportation options may be one of the biggest obstacles to its success. The good news is that public transit offers some faster options that undoubtedly generate less stress and pollution than driving.

The best thing to happen to encourage public transit usage has been high gas prices, and that's translated into ridership gains for such widely scattered systems as Utah Transit in Salt Lake City and Metrorail in Washington, D.C. In Canada, ridership has risen as much as 10 percent in cities like Vancouver and Winnipeg in step with rising gas prices, though cars remain the travel option of choice in the country's eastern cities.

According to the American Public Transportation Association, fourteen million Americans use one or another form of public transportation every weekday, while about seventeen million people drive their cars instead. The organization estimates that public transit ridership has grown by as much as 22 percent—faster than highway or air travel—since 1995. And a recently conducted Harris Poll concluded that the American public would like to see rail-based public transit "have an increasing share of passenger transportation."

Canadians have embraced public transit even more than their neighbors to the south. An estimated twelve million Canadians—including more than a fifth of all commuters in Toronto—use some form of public transit. Transportation analyst Paul Schimek found that public transit use is almost twice as high per capita in Canada than in the United States. And car use in Canada is almost 20 percent lower. Schimek attributes the differences to traditionally higher gas prices as well as to more compact urban development.

The strength of the American "highway lobby" is one reason Americans have been slow to embrace public transit. The powerful lobby, representing pavers and contractors, has worked directly with lawmakers over the years to encourage road building and private automobile use to achieve, in the words of a General Motors ad of days gone by, the "American dream of freedom on wheels."

CONTACTS: American Public Transportation Association, www.apta.com; Canadian Urban Transit Association, www.cutaactu.ca.

I visited New York City recently and could not believe the number of taxicabs on the streets. Are there any efforts to "green up" these vehicles?

Justin Grant, Berkeley, CA

Mayor Michael Bloomberg is pushing an ambitious plan to switch over the city's thirteen-thousand-vehicle taxicab fleet from gas-guzzling traditional cars to (comparatively) fuel-sipping gasoline-electric hybrids.

The idea is, from a base of one thousand green taxis at the end of 2008, to switch over 20 percent of the rest of the fleet to hybrids each year after. It's a really timely plan. Taxis spend much time idling in traffic and while waiting to load passengers. Hybrid cars, which pair a conventional gas engine with an electric motor, essentially shut

down when they are idling, minimizing emissions significantly. New York can reduce its emissions by more than 215,000 tons yearly just through taxis.

And even though cabbies will have to pay a premium to replace their existing vehicles with hybrids, most are behind the move, as it will save them about ten thousand dollars yearly in fuel costs alone. The Ford Crown Victorias that traditionally made up 90 percent of the taxi fleet get only ten to fifteen miles per gallon in city traffic. Ford's own Escape hybrid would improve that to thirty-four miles per gallon.

"I have been wanting to drive a hybrid taxi for years now," says Kwame Corsi, a cabbie from the Bronx. "Once this law allows us to drive hybrids, our gas mileage will skyrocket and our expenses will plummet. We pollute less and make more money—who can argue against that?" New York cabbies now ready to take the plunge can choose from any one of six different hybrid models, including the Ford Escape, the Toyota Prius and Highlander, the Lexus RX 400H, and the Honda Accord and Civic.

New York is not the first city to go hybrid with its cab fleet. San Francisco took the plunge in 2005 when forty Ford Escape hybrid taxis hit the streets there. The city's goal is to have half its taxi fleet—some six-hundred vehicles—powered by cleaner energy sources (either hybrids or compressed natural gas) by 2008.

Chicago's Carriage Cab Company is making the green switch, and the city has ordered taxi firms with over fifty cabs to add at least one hybrid to their fleets. The cities of Denver, Colorado, and Boston, Massachusetts, are also looking to make the transition.

Hybrid taxis have been plying the streets of Vancouver, British Columbia, since 2000, when cabbie Andrew Grant first started offering taxi rides in his Toyota Prius. Local lawmakers recently announced that the city would approve only eco-friendly vehicles when handling applications for new taxi companies or additions to existing fleets.

CONTACTS: Andrew Grant's Hybrid Taxi Driver blog, http://news.carjunky.com/alternative_fuel_vehicles/hybrid-taxi-cars-cde820.shtml; New York City Taxi and Limousine Commission, www.nyc.gov/html/tlc/html/home/home.shtml.

What are the health and environmental issues associated with noise and air pollution at airports?

John Cermak, via e-mail

On a 1997 questionnaire distributed to two groups—one living near a major airport and the other in a quiet neighborhood—two-thirds of those living near the airport indicated they were bothered by aircraft noise, and most said that it interfered with their daily activities. The same two-thirds complained more than the other group of sleep difficulties and also perceived themselves as being in poorer health.

Perhaps even more alarming, the European Commission considers living near an airport to be a risk factor for coronary heart disease and stroke, as increased blood pressure from noise pollution can trigger these more serious maladies. The commission estimates that 20 percent of Europe's population—about eighty million people—is exposed to airport noise levels it considers unhealthy and unacceptable.

Airport noise can also have negative effects on children's health and development. A 1980 study examining the impact of airport noise on children's health found raised blood pressure in kids living near Los Angeles's LAX airport. A 1995 German study found a link between chronic noise exposure from Munich's international airport and elevated nervous system activity and cardiovascular levels in children living nearby. And a 2005 study published in the prestigious British medical journal the *Lancet* found that kids living near airports in Britain, Holland, and Spain lagged behind their classmates in reading by two months for every five-decibel increase above average noise levels in their surroundings. The study also associated aircraft noise with lowered reading comprehension, even after socioeconomic differences were considered.

Living near an airport also means facing significant exposure to air pollution. Jack Saporito of the U.S. Citizens Aviation Watch Association (CAW) cites several studies linking pollutants common around airports—such as diesel exhaust, carbon monoxide, and leaked chemicals—to cancer, asthma, liver damage, lung disease, lymphoma, myeloid leukemia, and even depression. CAW is lobbying for the cleanup of jet engine exhaust as well as the scrapping or modification of airport expansion plans across the country.

Another group working on this issue is Chicago's Alliance of Residents Concerning O'Hare, which tries to rein in expansion plans at the world's busiest airport. According to the group, five million area residents may be suffering adverse health effects as a result of O'Hare, only one of four major airports in the region.

The Natural Resources Defense Council says that "airport air pollution is similar in scope to that generated by local power plants, incinerators, and refineries, yet is exempt from rules other industrial polluters must follow." Major airports rank among the top ten industrial air polluters in cities such as Los Angeles, Washington, D.C., and Chicago. The hundreds of thousands of airplanes taking off, landing, taxiing, and idling each day across the country emit contaminants into the air and ground that have been linked to a wide range of human health problems, including asthma and cancer.

And a 1999 report by the Intergovernmental Panel on Climate Change found that aircraft are responsible for 3.5 percent of greenhouse gas emissions worldwide; this could increase to 10 percent by 2050 as the popularity of air travel rises. And contrails, the vapor condensation trails you see overhead that are formed when airplanes fly at high altitudes through extremely cold air, could be contributing to global warming as they turn into high, thin cirrus clouds and trap heat from incoming sunlight within the atmosphere.

A recent agreement to cut thirty-seven daily peak-hour arrivals at Chicago's O'Hare should not only help to ease congestion and reduce delays but also to improve local air quality and reduce greenhouse gas emissions. Unfortunately, because of the increasing popularity of air travel, sixty of the hundred largest U.S. airports are proposing building more runways, thus expanding rather than reducing activity.

CONTACTS: Alliance of Residents Concerning O'Hare, www.areco.org; U.S. Citizens Aviation Watch Association, www.us-caw.org.

What is the status of bicycle use in the United States compared to other parts of the world like, say, China or Europe?

Monica Schmid, Seattle, WA

Given different types of weather and terrain—as well as historical economic and developmental trends—comparing bicycle usage in different parts of the world is tricky. What is clear, however, is that China dominates the world bike scene: a whopping 60 percent of the world's 1.6 billion bicycles are used daily by some five hundred million riders in China, who choose bikes over other modes of transport over half the time.

In Europe's hotbed of commuter bicycling, Amsterdam, residents choose their bikes 28 percent of the time, according to the International Bicycle Fund (IBF). In other European cities, the stats are also impressive: commuters choose bikes 20 percent of the time in Denmark, 10 percent in Germany, 8 percent in the United Kingdom, and 5 percent in both France and Italy. In stark contrast, the IBF reports that American city dwellers choose bikes less than 1 percent of the time. Meanwhile, estimates of the number of American adults who commute by bicycle regularly range from a low of four hundred thousand (based on U.S. census data) to a high of five million (according to the Bicycle Institute of America).

Unlike their American counterparts, Europe's urban planners are working to increase bicycle ridership. Copenhagen, for example, has three thousand bicycles available for short-term use for a small fee. Amsterdam provides covered bike parking at bus stops, encouraging both bike riding and mass transit at the same time.

In Muenster, Germany, bus lanes can be used by bikes but not by cars. Special lanes near intersections feed cyclists to a stop area ahead of cars, and an advance green light for cyclists ensures that they get through the intersection before cars behind them begin to move. Thanks to government programs to ease traffic congestion in Germany, bicycle use has increased by 50 percent over the past twenty years. Meanwhile, the United Kingdom has developed a plan to quadruple bicycle use by the year 2012. And in the European Union, bicycles have been included for the first time in the comprehensive transportation plan.

"European cities are much less suited to motoring and much more suited to short-distance bicycle transportation than are American cities," says transportation analyst John Forester. He cites historical reasons, including that European capitals were designed as walking cities served by rail, while America instead embraced cars.

Unfortunately for the world's air quality, a similar trend is developing in China, where people are ever more turning to cars and abandoning their bikes. Beijing, for instance, has been converting hundreds of bike lanes into car lanes and parking areas, as a recent influx of motor vehicles is maxing out existing roads. And with increased car traffic and fewer bike lanes, bicycle riding is getting more hazardous. "Nowadays there are just too many accidents, with a lot of cyclists getting hurt," says Zhang Lihua of the China Cycling Association. "Riding bicycles is becoming too inconvenient and too dangerous," he adds.

CONTACTS: Earth Policy Institute, www.earth-policy.org; International Bicycle Fund, www.ibike.org.

What exactly is eco-travel or eco-tourism?

Jeannette Peclet, Norwalk, CT

The International Ecotourism Society (TIES) defines eco-tourism as "travel to natural destinations that minimizes impact, builds environmental awareness, helps fund conservation, and respects and sustains local cultures while supporting human rights and democracy."

The United Nations Environment Programme (UNEP) sees it as travel focused on "the observation and appreciation of nature as well as the traditional cultures prevailing in natural areas." UNEP says eco-tours must have educational features, be organized for small groups by locally owned businesses, minimize negative impacts "upon the natural and socio-cultural environment," and support the protection of natural areas by generating income for the host communities to use in conserving and sustaining their natural and cultural resources.

Recent studies indicate that as much as 7 percent of all tourism worldwide operates under some sort of "eco" label. One recent survey concluded that eight million U.S. travelers have taken at least one "eco-tourist" holiday, while another concluded that three-quarters of all Americans have taken a trip involving nature and the outdoors. In the Asia-Pacific region, eco-tourism accounts for 20 percent of all travel. Meanwhile, in Africa, where most visitors travel to nature reserves and game parks, the figures are even higher. The Kenya Wildlife Service, for instance, estimates that 80 percent of visitors come to see wildlife.

But the debate over what types of travel constitute eco-tourism has meant that a wide range of dining, lodging, and transportation vendors advertise themselves as "green" regardless of whether their operations meet the criteria defined by TIES and other groups. As *E* describes it, "A beachfront hotel tower built of imported materials with absentee owners and no local employees is not an eco-resort, even if it does offer its guests the option of not washing their towels."

And travelers should keep in mind that "adventure" travel or "nature-based" tourism trips are not necessarily environmentally friendly. In fact, tour operators offering access to remote scenic and wild locations need to take extra care so that their trips do not endanger the very flora, fauna, and geological features they are offering to showcase. Sad stories of so-called eco-tourism run amok—where overvisitation has led to trampled landscapes and damaged wildlife habitat—abound from the Galápagos Islands and Mexico's Chiapas region to the coastal caves of Thailand, the reefs of Hawaii, and beyond.

If you want authentic eco-tourism, you need to ask travel vendors a lot of questions about how they operate. That's the only way to know if they're harming or helping local environments and cultures.

CONTACTS: The International Ecotourism Society, www.ecotourism.org; UN Environment Programme, www.unep.org.

Many hotels now ask guests to reuse towels and not request new sheets every day, to save water and energy. What other eco-friendly trends, if any, are taking place in the hospitality industry?

Jenny Baker, Bozeman, MT

Asking guests to reuse towels and keep their sheets for more than one night are just a few of the many ways the lodging industry has been rapidly "greening up" operations in recent years. According to the Green Hotels Association (GHA), a trade group, thousands of member establishments share common goals of saving water, energy, and other resources—while saving money—to help protect the planet.

And travelers don't seem to mind. Such minor concessions as reusing towels and sheets are garnering some 70 percent participation among guests, according to GHA, and in the process are saving lodging establishments as much as 5 percent on utility bills.

Other examples of green-friendly hospitality strategies abound. Shower wall-mounted body washes (ubiquitous in Europe) are replacing tiny, individual soap bars and disposable shampoo and conditioner bottles. Room lights are being retrofitted with energy-saving compact fluorescent bulbs in lieu of incandescent energy hogs. Some establishments are installing "occupancy sensors" so that lights go on and off automatically when guests enter or leave their rooms. Still others are providing cloth laundry bags made from retired sheets and installing "low-flush" toilets to save water.

Hotel restaurants and banquet facilities are also getting in on the act, serving water upon request only, shunning disposable paper and plastic in favor of reusable pitchers and pourers for cream and sugar, and using small serving dishes in place of single-serving pouches for butter and jellies. Some are also using coins or chips for car parking and coat checking instead of paper tickets. Outdoors, solar energy is powering signs and, in tropical areas, heating water. And mowed landscaping is being replaced by plantings and other kinds of ground cover to reduce lawn mower use and its inherent air and noise pollution.

Green Seal, a nonprofit organization that promotes the use of environmentally friendly products and practices, began working with the lodging industry a decade ago. The group works to educate the fifty-four thousand U.S. hotels and motels about the economic benefits of environmental choices. It also publishes *Greening Your Property*, a comprehensive guide to green purchasing and operations for the hospitality industry. The guide helps hoteliers find environmentally friendly options for nearly everything they buy—from industrial cleaners and floor care products to paints, lighting, and recycled paper towels, tissues, and napkins. Additionally, Green Seal's hotel certification program helps tourists, meeting planners, and business travelers identify environmentally responsible lodging options.

The green trend has taken off at smaller independent hotels, but the large chains are paying attention too. Placards asking guests to reuse towels are already ubiquitous tubside at hotels operated by the likes of Best Western, Marriott, Ramada, Sheraton, and Westin, to name a few. And many of these chains have enacted far-reaching practices and policies involving environmental sustainability.

CONTACTS: Green Hotels Association, www.greenhotels.com; Green Seal, www.greenseal.org.

There has been so much attention paid to designing environmentally friendly cars. Is there a similar effort to replace gas-guzzling boats?

Margot Fischer, via e-mail

We've been regulating fuel economy and emissions in cars and trucks for decades but have gotten a late start addressing similar issues with boats. In 1996, though, recognizing a growing problem of boat engine pollution, the EPA issued new rules to "bring forth a new generation of marine engines featuring cleaner technology and providing better engine performance to boat owners."

Even small quantities of fuel and exhaust discharged by boats can disrupt freshwater and marine ecosystems. The cumulative effect of millions of inefficient motorboats plying our waterways has been devastating to marine life and our water supplies. Under the new EPA regulations, which will phase in over the next thirty years, new marine engines will burn gas much more efficiently and generate much less pollution than most models out on the water today.

According to the EPA, traditional two-stroke boat engines waste significant amounts of gasoline and oil, spilling as much as 30 percent of their fuel into the water and air either unburned or partially unburned. In the water, unburned hydrocarbons increase concentrations of benzene, methyl tertiary-butyl ether (MBTE) and other toxic substances that pollute water ecosystems. In the air, they help form smog, which causes a host of health problems and disrupts visibility everywhere from our cities to our national parks.

If you're thinking of buying a boat, you should look for either a four-stroke or direct fuel injection (DFI) two-stroke engine. These pollute about 75 percent less than their traditional two-stroke predecessors and use as much as 50 percent less gas and oil. They cost more than traditional two-stroke engines, but owners soon make up the difference in fuel and oil savings. They are also easier to start and maintain and are quieter.

New generations of electric boat motors are also coming into use, and promise to significantly cut pollution if adopted widely. Wooden, sport, and leisure boats are now all available with electric engines that are quite comparable to traditional engines in performance and looks. They are also nonpolluting, quiet, and can cruise where gas motors are not permitted. Some leading makers include Beckman, Budsin, Cobalt Marine, Electric Launch, Duffy, Elco ElectraCraft, Griffin Leisure, Pender Harbour, and Spincraft.

The only catch is that the energy that powers electric boats most likely comes from a coal-burning power plant. A handful of manufacturers–including Australia's Solar Sailor and Canada's Tamarack Lake–make solar-powered or solar-assisted electric boats.

Of course, the ultimate energy source for any recreational activity is elbow grease. But if you need more than a canoe or kayak to get around, Nauticraft hybrid boats employ human pedal power to augment a small electric motor. And the Italian-made Shuttle Bike puts a new spin on pedal boats: owners affix two inflatable pontoons to their mountain bikes, and they can then pedal around their local lake or harbor.

CONTACT: EPA's *Shipshape Shores and Waters: A Handbook for Marina Operators and Recreational Boaters*, www.epa.gov/owow/nps/marinashdbk2003.pdf.

What is the most environmentally friendly way I can wash my car: doing it myself or going to the local car wash?

Jim S., Denton, TX

Washing cars in our driveways is one of the most environmentally *unfriendly* choices we can make. Unlike household wastewater that enters sewers or septic systems and undergoes treatment before it is discharged into the environment, what runs off from your car goes right into storm drains and eventually into rivers, streams, creeks, and wetlands, where it poisons aquatic life and wreaks ecosystem havoc. The water running off your car is contaminated with a witches' brew of gasoline, oil, and residues from exhaust fumes as well as the harsh detergents being used for the washing itself.

Federal laws in both the United States and Canada require commercial car-wash facilities to drain their wastewater into sewer systems, so it gets treated before it is discharged back into the great outdoors. And commercial car washes use computer-controlled systems and high-pressure nozzles and pumps that minimize water usage. Many also recycle and reuse the rinse water.

The International Carwash Association, an industry group representing commercial car-wash companies, reports that automatic car washes use less than half the water of even the most careful home car washer. According to one report, washing a car at home typically uses between 80 and 140 gallons of water, while a commercial car wash averages less than 45 gallons per car.

If you must wash your car at home, choose a biodegradable soap specifically formulated for automotive parts, such as Simple Green's Car Wash or Gliptone's Wash 'N Glow. Or you can make your own biodegradable car wash by mixing 1 cup of liquid dishwashing detergent and 3/4 cup of powdered laundry detergent (each should be chlorine and phosphate free and nonpetroleum based) with 3 gallons of water. This concentrate can then be used sparingly with water over exterior car surfaces.

Even when using green-friendly cleaners, it is better to avoid the driveway and instead wash your car on your lawn or over dirt so that the toxic wastewater can be absorbed and neutralized in soil instead of flowing directly into storm drains or open water bodies. Also, try to sop up or disperse those sudsy puddles that remain after you're done. They contain toxic residues and can tempt thirsty animals.

One way to avoid such problems altogether is to wash your car with waterless formulas, which are especially handy for spot cleaning and are applied via spray bottle and then wiped off with a cloth. Freedom Waterless Car Wash is a leading product.

One last caution: Kids and parents planning a fund-raising car-wash event should know that they might be violating clean water laws if runoff is not contained and disposed of properly. Washington's Puget Sound Car Wash Association, for one, allows fund-raisers to sell tickets redeemable at local car washes, enabling the organizations to make money while keeping dry and keeping local waterways clean.

CONTACTS: Freedom Waterless Car Wash, www.freedomwaterlesscarwash.com; International Carwash Association, www.carcarecentral.com; Puget Sound Car Wash Association, www.charitycarwash.org; Simple Green, www.simplegreen.com.

10
TECHNICALLY SPEAKING

Saving the Earth, One iPod at a Time

Are you listening? Chances are, you're listening, typing, watching, talking, and "gaming" simultaneously. We guzzle information and entertainment like we guzzle gas, and our electronics habit is having its own serious environmental impact. For as great as our technological advances are, they often lead us to buy a new cell phone or computer or television or camera every couple of years—or less—leaving huge amounts of toxic waste behind with each slight innovation. But while technology poses health concerns (like the amount of radioactive waves we're exposed to), it also provides us with solutions. Educated consumers are already demanding more energy-efficient laptops and less wasteful packaging on tiny items like memory cards and chargers. All the accessories that go with our techie devices—the batteries and cartridges—can be recycled too. And our national shift to downloadable everything is already impacting the amount of waste we produce and energy we consume, replacing DVDs and CDs and all the packaging and shipping they require—which, to borrow from an old phrase, is something to sing, text, and blog about.

What are the environmental implications of the proliferation of iPods specifically and digital music in general?

Mike Romano, San Francisco, CA

The downloading of digital music and other forms of entertainment from the Internet has staggering repercussions not only for the music industry and the consumer experience but also for our beleaguered planet.

Analysts estimate that American consumers buy about a billion CDs every year, most of which eventually end up in landfills or incinerators. Songs downloaded from the Internet consume only hard drive space and don't contribute directly to the waste stream. To get rid of downloaded music, one need only drag it to the trash symbol on the desktop. As far back as 2006, consumers had already downloaded more than a billion songs via Apple's iTunes service alone. If all that music had been copied to CDs it would have filled up eighty-five million disks.

That's not the whole story, of course. Downloaded music has to be played, and a large amount of "e-waste" (electronic waste) is already clogging landfills in every community. Consumer electronics contain large amounts of nasty toxins such as lead, mercury, and cadmium, so when computers, monitors, and MP3 players end up in landfills, they can contaminate the surroundings and become a serious health issue.

iPod maker Apple has come under fire from the Silicon Valley Toxics Coalition. The group charges that Apple's hardware recycling program, which accepted fifteen hundred tons of old equipment from consumers last year, is woefully inadequate and that Apple has been lobbying against legislation that sets up such "take-back" programs. The group also claims that iPod batteries wear out too soon, prompting consumers to junk the gadgets prematurely. Apple disputes the charges, claiming it is one of the most environmentally responsible businesses in the industry.

But despite all this, digitally downloaded music still seems to be the more environmentally friendly choice, especially if consumers make efforts to recycle as much of their e-waste as possible. Apple, Dell, HP, and IBM all offer low-cost ways to recycle hardware directly.

Meanwhile, some groundbreaking new CDs, one made from corn and another partly from paper, are on the horizon. Sanyo has teamed up with NatureWorks (formerly Cargill Dow) in the production of MildDiscs made from corn (one ear of corn can make ten CDs). And Sony's Blue-ray discs are 51 percent paper, can be cut with scissors, and will hold about half the data of a computer hard drive.

CONTACTS: Apple product take back, www.apple.com/environment/recycling; Silicon Valley Toxics Coalition, www.svtc.org.

As an online gamer, I spend a lot of time in front of a computer. What's the environmental impact? And are "greener" PCs available?

Bob Grant, Burlington, VT

Online gamers and other big-time computer users are definitely leaving an environmental mark. Depending on when they were made and how they were designed, standard desktop PCs can use anywhere from 60 to 300 watts when in use, while an inefficient gaming PC with a powerful graphics card, multiple hard drives and optical drives, flash memory reader, and thirty-inch LCD might consume as much as 750 watts. That's as much as your typical refrigerator! Until 2007, government Energy Star requirements only measured a computer's energy use while in standby mode, which allowed the majority of brands to carry the label.

The stricter efficiency requirements for computers have brought "greener" models from manufacturers. You'll find the largest selection from companies like Dell and Hewlett-Packard, which sell preconfigured PCs and notebooks. Many businesses use the Electronic Product Environmental Assessment Tool (EPEAT) to assist in the purchase of "greener" computing systems, and the evaluations can be useful to consumers too. The council evaluates computing equipment on twenty-eight efficiency and sustainability criteria and rates them as bronze, silver, or gold for energy usage.

Technology company VIA is well regarded as an industry leader in low-wattage central processing units (CPUs), with some barely sipping only a dozen or so watts from the power supply. Some typical VIA designs can outperform competitors using only 23 watts, or less than half the power called for by Energy Star specifications. Of course, graphics cards used by PC gamers are serious energy hogs. Your top-end ATI or Nvidia card will render great graphics, but gobble up 300 watts or more. Newer cards are better, but much depends on their use. The best advice is to save by buying only the graphics power you need.

One of the easiest ways to save on computer power is to use technology that automatically rests when you do, and to shut down when you're not using your computer, a standard feature in new computer models. Windows XP allows for the user to configure power management settings and Vista Ultimate lets you configure power-saving options in even more ways. Vista can actually throttle its power consumption for some tasks, saving energy while you work. If you are just typing a document in Microsoft Word, performance will back down, whereas, if you are editing video in a powerful program like Adobe Premier Pro, Vista will use all the processing power available.

Screen savers are not energy savers, and power-down features may not work if you have a screen saver activated. Happily, LCD color monitors do not need screen savers at all. In terms of shutting down, while PCs use a small amount of energy when they start up, it's considerably less than the energy used when a computer is running for long periods of time. Consider turning off the monitor if you aren't going to use your PC for more than twenty minutes and both the CPU and monitor if you're not going to use your PC for more than two hours.

Don't be concerned about the "wear and tear" of turning PCs on and off. Most PCs reach the end of their "useful" life due to advances in technology long before the effects of being switched on and off can impact their service life.

CONTACTS: Electronic Product Environment Assessment Tool, http://epeat.net; Energy Star www.energystar.gov; recycling an old monitor, www.epa.gov/epaoswer/hazwaste/recycle/ecycling/donate.htm; VIA, www.via.com.

How can I recycle my outdated computer equipment?

Kenneth Rapp, Toms River, NJ

These days, you can expect a new computer to serve you for three to five years at best before "must-have" features become available only in newer models. Many companies have "computer graveyards," rooms filled to the ceiling with outdated computers, printers, monitors, cables, and other accessories that are no longer in operation and seemingly have nowhere to go but the junk heap.

It's no surprise, then, that more than ten million computers end up in American landfills every year. But old computer equipment languishing in landfills poses myriad environmental hazards, as many contain toxic compounds that can seep into surrounding land and groundwater. To protect users against radiation, the average PC monitor contains five pounds of lead. Cadmium, mercury, and chromium are among the dangerous metals found in circuit boards, and plastic housings are doused with toxic flame retardants.

Nikki and David Goldbeck's book *Choose to Reuse* reports that many computers can be saved and don't need to end up in landfills. The first thing to check is if your old computer can be upgraded; often the substitution of a simple memory chip can make a slowpoke speed up considerably. And RAM memory—provided there are sufficient expansion slots—is getting cheaper all the time.

If an upgrade won't work, there are alternatives to landfills. Goodwill and the Salvation Army will take working older equipment and resell it. "Free Computer" ads can be posted at schools and workplaces. And brokers like American Computer Exchange will take your hardware for trade on a newer model.

Many worthy nonprofit groups will make good use of outdated computer equipment. The National Cristina Foundation, for instance, places used technology with nonprofit organizations and public agencies that serve the poor and disabled. For a more do-it-yourself approach, the Global Crisis Solution Center provides a free online resource hooking up equipment donors with needy nonprofits.

Europe is leading the way in keeping computers out of landfills, with all computer manufacturers required to have recycling programs in place. In the United States, several makers will now recycle or exchange computers, often for a marginal fee. IBM, Dell, and Hewlett-Packard have all started such programs.

CONTACTS: American Computer Exchange, www.amcoex.com; Dell recycling, www.dell.com/content/topics/segtopic.aspx/dell_recycling?c=us&cs=19&l=en&s=dhs; Hewlett-Packard recycling, www.hp.com/hpinfo/globalcitizenship/environment/recycle; Global Crisis Solution Center, www.globalcrisis.info/computerrecycle.html; IBM recycling, www.ibm.com/ibm/environment; National Cristina Foundation, www.cristina.org.

Is it economical and environmentally friendly for me to recycle my empty inkjet printer cartridges instead of buying new ones?

Matt Hoffman, Seattle, WA

More than three hundred million inkjet printer cartridges find their way into American landfills every year. And believe it or not, each one requires about three quarts of oil and other raw materials to produce and also contributes its fair share of greenhouse gases during manufacturing. As anyone who has ever bought one knows, they come packaged in excessive amounts of cardboard and plastic, so it often takes several minutes and a pair of strong scissors to break through and get to the cartridge!

Any effort to reuse or recycle cartridges is a big win for the environment. And given the exorbitant prices of new inkjet cartridges—the real profit center for printer manufacturers—it also makes economic sense for consumers who just want to save money.

The good news is that Americans are already recycling more than forty thousand tons of inkjet cartridges each year. Hundreds of companies are eager to pay for your used cartridges so they can reink them and resell them at prices much lower than new ones.

Webuyempties.com, inkjetcartridge.com, and the eCycle Group, among others, take back major-brand inkjet printer cartridges and pay for the privilege, even reimbursing shipping costs. These companies only accept large quantities (like one hundred or more) of spent cartridges, paying between ten cents and five dollars each, depending on the cartridge type. And Staples, Office Depot, and OfficeMax each give customers about three dollars in store credit, or in some cases a ream of office paper, for each empty cartridge returned.

Most of the major inkjet-printer manufacturers, including Hewlett-Packard, Epson, Canon, and Lexmark, will gladly take empty cartridges shipped back in their original boxes. Hewlett-Packard helpfully puts prepaid return shipping labels inside its boxes to facilitate customer recycling of its used inkjet cartridges.

Several companies offer special buyback rates for schools, churches, and other nonprofits, which can solicit and collect used cartridges from members and businesses to raise money. Interested organizations can contact companies like iRethink and FundingFactory, which both have special programs to facilitate collection and reimbursement for spent inkjet cartridges.

If you don't mind getting your hands a little messy, you can reink your empty cartridges yourself. Squeeze-bottle ink refills are the most cost-effective and environmentally friendly way to keep on printing. Inkjetman, which sells its own refilled inkjet cartridges, also sells inkjet refill kits, which will last through thousands of pages, for about the price of a single new cartridge. Fill Jet sells similar kits and estimates the cost of a refilled cartridge to be about two dollars in ink, which represents a savings of at least 80 percent over buying refilled and recycled cartridges from them.

Keep in mind, however, that *Consumer Reports* has tested refill kits and says that most are "messy," with the exception of the relatively easy-to-use Automatic Refill System from Dataproducts.

CONTACTS: Dataproducts, www.dpc.com; eCycle Group, www.ecyclegroup.com; FundingFactory, www.fundingfactory.com; iRethink, www.irethink.com; Webuyempties, www.webuyempties.com.

Is it true that computers and other electronic devices contain arsenic and other toxins, and if so should I worry about using them?

Jen Deeds, McLean, VA

As any murder mystery enthusiast knows, arsenic can be lethal if ingested in large amounts. Electronics manufacturers use it as an efficient conductor of electricity; useful when periodic strong bursts are needed. But don't worry—the traces of the naturally occurring element that can be found inside your calculator, watch display, television set, or computer are too small to hurt you directly.

But the toxins in electronics do pose community-wide dangers if not disposed of properly. Recent research shows that many common electronic devices qualify as hazardous materials according to existing EPA definitions due to the arsenic, mercury, and lead within. They should only be discarded in permitted hazardous-waste-treatment facilities.

Unfortunately, though, many of these discarded products will end up in landfills not equipped to handle hazardous waste, and their arsenic and other toxins can make their way into groundwater. The resulting drinking water contamination has been linked to a wide range of human ailments, including bronchitis, liver cirrhosis, and even some cancers. In fact, the EPA considers arsenic to be a carcinogen.

A Silicon Valley Toxics Coalition report estimates that five hundred million computers—not to mention many more millions of televisions, calculators, and MP3 players—become obsolete annually. While there have been no studies on arsenic specifically, researchers have found that about 40 percent of the toxic lead found in U.S. landfills in recent decades originated with discarded electronics. Further, as much as 80 percent of U.S. electronic waste collected for recycling today is sent to China, India, and Pakistan, so the computer you abandon today could end up contaminating the drinking water in a developing country tomorrow.

The best alternative to adding to the waste stream is to upgrade or repair your old computer or TV to keep it humming along happily at home or in the office—and out of any landfill near or far. By keeping your vintage electronics around, you help safeguard your community and others from toxic waste.

But if the old gear really has to go, the Seattle-based Basel Action Network lists by region electronics-recycling companies that adhere to high standards with regard to both environmental and health considerations. In addition, American and Canadian consumers can look for products that are also sold in Europe, as manufacturers who sell there must by law avoid using toxins like arsenic and lead. And if your old model still works at all, it may be a candidate for a donation to a local school or through Gifts in Kind, a clearinghouse for usable used stuff. Lastly, some computer makers, including IBM and Hewlett-Packard, have in-house programs to take back and recycle old models.

CONTACTS: Basel Action Network, www.ban.org/pledge/locations.html; EPA, "Arsenic Compounds" page, www.epa.gov/ttn/atw/hlthef/arsenic.html; Gifts in Kind, www.giftsinkind.org; Hewlett-Packard recycling, www.hp.com/hpinfo/globalcitizenship/environment/recycle; IBM recycling, www.ibm.com/ibm/environment; Silicon Valley Toxics Coalition, www.svtc.org.

What happens to my old cell phone after I upgrade? Do the stores really recycle them or give them to the poor, or are they just ending up in landfills? Where can I take mine to ensure that it is dealt with properly?

Paul G., Reno, NV

Cell phones are giving computers and monitors competition for the dubious distinction number one contributor to the world's e-waste problem. Toxin-laden electronics are clogging landfills and polluting air and groundwater supplies from coast to coast.

The average North American gets a new cell phone every eighteen to twenty-four months, and that makes the old ones—which contain hazardous materials like lead, mercury, cadmium, brominated flame retardants, and arsenic—the fastest-growing type of manufactured garbage in the nation. According to the EPA, Americans discard 125 million phones each year, creating sixty-five thousand tons of waste.

The good news is that electronics recyclers are stepping in to help. Nonprofit Call2Recycle has a website that lets consumers enter a zip code and be directed to a drop box in their area. Most major electronics retailers, from Radio Shack to Office Depot, participate in the program and offer Call2Recycle drop boxes in their stores. Call2Recycle recovers the phones and sells them back to manufacturers, which either refurbish and resell them or recycle their parts for use in making new products.

CollectiveGood takes used cell phones, refurbishes them, and then resells them to distributors and carriers for use primarily in developing countries, providing affordable communications to poorer citizens while helping to "bridge the digital divide." The group also recycles all nonfunctioning batteries through a partnership with the Rechargeable Battery Recycling Corporation. When you donate your phone to CollectiveGood you can direct the profits from the sales to a charity of your choice.

Another player is ReCellular, which manages the in-store collection programs for Bell Mobility, Sprint PCS, T-Mobile, Best Buy, and Verizon. The company also maintains partnerships with Easter Seals, the March of Dimes, Goodwill Industries, and other nonprofits that collect cell phones to fund their charitable work. According to ReCellular vice president Mike Newman, the company is trying to get consumers to "automatically think of recycling cell phones, just like they currently do with paper, plastic, or glass."

Neither the United States nor Canada mandates electronics recycling of any kind at the federal level, but a few states and provinces are getting into the act on their own initiative. California recently passed the first cell phone recycling law in North America. Since 2006, electronics retailers doing business there must have a cell phone recycling system in place in order to legally sell their products, whether online or in-store. Other U.S. states and several Canadian provinces are considering legislation.

CONTACTS: Call2Recycle, www.rbrc.org/call2recycle; CollectiveGood, www.collectivegood.com; ReCellular, wirelessrecycling.com/home.

Could our health be negatively affected by all the radio and microwave radiation emitted by cell phones, cell phone towers, and wireless pagers, among other things?

Beverly Filip, Santa Cruz, CA

Cell phones certainly do emit low levels of electromagnetic radiation. While it is widely known that sufficient levels of nonionizing radiation heat up body tissue and increase the risk for tumor growth, no conclusive link between cell phones and cancer has been found. "It's difficult to collect reliable data on the potential harm caused by cell phone use because the devices are so new," says Libby Kelley, executive director of the Council on Wireless Technology Impacts.

A study of 420,000 cell phone users in Denmark concluded that there is no link "between the use of [cellular] phones and brain tumors and cancers of the brain or salivary gland or leukemia." Researchers noted that a typical cell phone functions at a low power level, resulting in "a very low rise in brain temperature." However, Swedish scientist Lennart Hardell had made a link between brain cancer and older analog cell phones used for at least eight years.

Some earlier studies suggested a link between exposure to radiation from cell phones and an increased risk of acoustic neuroma, a cancerous tumor of the nerve connecting the ear to the brain, but more recent research found no such links. The issue is primarily heat. According to the Occupational Safety and Health Department of the Communication Workers of America (CWA), "As high frequency radio frequency radiation . . . penetrates the body, the exposed molecules move about and collide with one another causing friction and, thus, heat . . . If the radiation is powerful enough, the tissue or skin will be heated or burned . . . There is substantial scientific data that establishes negative health effects associated with microwave radiation."

CWA cites cataracts as one possible negative health effect from prolonged exposure, as well as nervous system damage and even reproductive problems in both males and females. This issue was in the news in 1992 over the issue of the safety of police radar devices, but subsequent studies were inconclusive.

The results of a 2003 study do not bode well for habitual chatterers. Researchers documented brain damage in laboratory rats exposed to radio frequencies from cell phones at levels comparable to what people would experience during normal use. The study's authors expressed concern that "after some decades of (often) daily use, a whole generation of [cell phone] users may suffer negative effects, perhaps as early as middle age."

The environmental effects of radio frequencies are also largely unclear. Migrating birds have been known to fly right into cell phone and other communications towers. Some blame the radiation from the towers for disorienting birds and undermining their navigational abilities. Others chalk such incidents up to poor visibility associated with bad weather. Some farmers have observed that cows grazing near cell towers are more likely to experience still births, spontaneous abortions, birth deformities, and behavioral problems, not to mention general declines in overall health. Moving cattle herds away from the towers has reportedly led to immediate health improvements.

CONTACTS: Council on Wireless Technology Impacts, (415) 892-1863, www.energyfields.org; Department of Labor, "Radiofrequency and Microwave Radiation," www.osha.gov/SLTC/radiofrequency radiation.

Are digital cameras more environmentally friendly than traditional cameras?

Ann Veddern, Mason, OH

It may be counterintuitive, but your best option could be a single-use camera, often referred to as a disposable camera, according to James Blamphin, manager of environmental news and information at the Eastman Kodak Company. "The single-use camera has the highest recycling rate of any consumer product," he says. "We've turned a waste stream into a revenue stream." Since 1990, Kodak has recycled more than 750 million one-time-use cameras through its closed-loop recycling program. The polystyrene covers and viewfinders are ground down and reprocessed into new camera components, and lens acrylic is made into toothbrushes.

Digital cameras are also evolving to be more environmentally friendly, Blamphin explains. They're getting smaller, for one thing: the camera body mass has been reduced by 50 percent over the last five years, requiring fewer resources. Cameras now run on fewer batteries, and faster computer downloading technology reduces energy use.

Kodak and other companies, such as Canon, have recently removed lead from lenses. Kodak has also removed cadmium from sensors and mercury from displays.

Traditional cameras get a bad rap primarily because of the chemicals required for processing photos. But since 1968, the photo chemicals required to develop and print one twenty-four-exposure roll has been reduced from two quarts to three ounces—a 96 percent reduction, and chemicals, including the silver used, can be reused, says Blamphin. He stands by his nomination of the single-use camera as the most environmentally friendly option, however, because the recycling process saves resources. After seven or eight years, you're going to get another camera and discard the old one anyway; a single-use camera can be recycled up to ten times, he explains. And regarding digital technology: "Don't forget the chemistry needed for inkjet printing of digital images, and computer chip waste, which there are not well-established recycling programs for," he says.

CONTACT: Eastman Kodak Company, (800) 235-6325, www.kodak.com.

I've heard that it's now safe to throw away common household batteries and that only rechargeable batteries should be recycled. Is this true?

Doug Reynolds, Martinsville, IN

Today's common household batteries—the ubiquitous AA, AAA, C, D, and nine-volt batteries from Duracell, Eveready, and others—don't pose as great a threat to modern landfills as they used to because they contain much less mercury than their predecessors. For that reason, most towns and cities recommend simply throwing batteries away with your trash.

Still, many people will feel better recycling their batteries, as they still contain trace amounts of mercury and other potentially toxic stuff. Some municipalities will accept these batteries (as well as older, more toxic ones) at household hazardous-waste facilities, which will send them to be processed and recycled as components in new batteries.

Other options abound, such as the mail-order service Battery Solutions, which will recycle your spent batteries at a cost of eighty-five cents per pound. To find a company near you to drop off your old batteries for recycling, check out the comprehensive national database at the Earth911 website. Meanwhile, the national chain Batteries Plus is happy to take back disposable batteries for recycling at any of its 255 retail stores from coast to coast.

If you do come across batteries made before 1997—when Congress mandated a widespread mercury phaseout in batteries of all types—it's definitely worth the effort to recycle them, because they may contain as much as ten times the mercury in newer versions.

And spent rechargeable batteries from cell phones, MP3 players, and laptops are indeed a big issue. Although they have many environmental advantages, rechargeable batteries typically have toxic heavy metals sealed up inside, making them a threat to both landfills and incinerators. Luckily, the battery industry sponsors the Rechargeable Battery Recycling Corporation (RBRC), which facilitates the collection of used rechargeable batteries collected in an industry-wide "take-back" program for recycling.

Consumers can help by buying only electronics that carry the RBRC logo. You can find drop-off locations for rechargeable batteries and cell phones by calling RBRC's hotline at 1-800-8BATTERY or by visiting the online drop location finder at rbrc.org. Also, most Radio Shack stores will take back rechargeable batteries and deliver them to RBRC free of charge. RBRC then processes the batteries via a thermal recovery technology that reclaims metals such as nickel, iron, cadmium, lead, and cobalt, repurposing them for use in new batteries.

CONTACTS: Battery Solutions, www.batteryrecycling.com; Earth911, www.earth911.org; Batteries Plus, www.batteriesplus.com; Rechargeable Battery Recycling Corporation, www.rbrc.org.

Other than calculators, what are some accessories and gadgets that are now available solar-powered?

Frank Rogers, Concord, NH

Perhaps the most widespread use of energy from the sun today is for charging up small electronic devices like flashlights, watches, Palm-Pilots, and cell phones. Solar cells are also being put to use around the home to power garden, pool, and security lighting as well as automatic watering and lawn-feeding devices. And as photovoltaic technology improves, people are using small solar cells to power up bigger devices like radios, cameras, and even laptop computers. A good assortment of such items can be ordered from online stores such as Brunton, Sundance Solar, Real Goods, Global Merchant Imports, and Energy Federation (EFI).

In the developing world, EFI sells a Sun Oven solar cooker, which is reducing reliance on increasingly scarce and threatened wood products. The interior of the oven is heated by passive solar energy when the oven's reflectors are opened up and pointed toward the sun. And it's not just an oven; food can be baked, boiled, and steamed at temperatures of 360 to 400 degrees Fahrenheit.

Wired reports that climber Sean Burch used solar cells to charge his laptop and phone during his solo ascent of Mount Everest in 2003. "The sun was so bright at 18,000 feet that it wasn't a problem at all," says Burch, who didn't have the manpower to bring along the hundred-pound batteries used by bigger climbing crews to power communications devices. "It was nice because I had my computer, solar panels, and phone and I could communicate as well as anyone," he adds.

CONTACTS: Brunton, www.brunton.com; Energy Federation, www.efi.org; Global Merchant Imports, www.global-merchants.com; Real Goods, www.realgoods.com; Sundance Solar, www.sundancesolar.com.

NOTES

CHAPTER 1: EAT, DRINK, AND BE WARY

Page

3 “Farmers in India . . .”, Nadia El-Hage Scialabba and Caroline Hattam, eds., “Organic Agriculture, Environment and Food Security,” UN Food and Agriculture Organization, Rome, 2002.

3 “Even when they don’t get a bumper crop . . .” Cited by Dr. Liz Stockdale of Britain’s Institute of Arable Crop Research.

3 “Only slightly more than 2 percent of all farms in the United States are currently organic.” Figure cited by the Organic Consumers Association.

4 “Local food is often safer, too . . .” Center for a New American Dream, www.newdream.org/consumer/farmersmarkets.php.

6 “The majority of corn, soy, and cotton grown by American farmers . . .” According to the nonprofit Pew Initiative on Food and Biotechnology.

6 “When the Flavr Savr hit store shelves . . .” See *First Fruit: The Creation of the Flavr Savr Tomato and the Birth of Biotech Food* by Calgene researcher Belinda Martineau (New York: McGraw-Hill, 2001).

7 “Sugar may be responsible for more biodiversity loss than any other crop . . .” World Wildlife Fund, *Sugar and the Environment*, 2004. It’s online at http://assets.panda.org/downloads/sugarandtheenvironment_fidq.pdf.

11 “Americans consume some three hundred million cups of coffee every day.” According to the Specialty Coffee Association of America.

13 “. . . the nondairy organics did an incredible $757 million in sales.” Elizabeth

Furhman, "Modern Organics: On the Cutting Edge of Hip and Healthful," *Beverage Industry*, December 1, 2006.

16 "Just about every aspect of meat production . . . is an environmental disaster with wide and sometimes catastrophic consequences." Jim Motavalli, "So You're An Environmentalist: Why Are You Still Eating Meat?" *E – The Environmental Magazine*, January/February 2002. It's online at www.emagazine.com/view/?142.

16 "Livestock is a 'major player' in climate change . . ." United Nations report, *Livestock's Long Shadow,* 2006. It's online at www.virtualcentre.org/en/library/key_pub/longshad/A0701E00.htm.

16 "In 1950, world meat production was 44 million pounds annually . . ." Lester R. Brown, *Outgrowing the Earth: The Food Security Challenge in an Age of Falling Water Tables and Rising Temperatures* (New York: W. W. Norton, 2005).

18 "Nationwide, more than three thousand bodies of water were under fish-consumption advisories in 2003." EPA, *Clean Safe Water,* 2003. It's online at www.epa.gov/cfo/par/2003par/ar03_goal2.pdf.

22 "A number of major chain stores . . . stock only dolphin-safe tuna." List of chain stores and restaurants from Defenders of Wildlife.

29 "Urban gardens, like the ones springing up all over New York City and Seattle, provide 15 percent of the world's food supply." Figure cited by the United Nations Development Programme.

CHAPTER 2: THE ENLIGHTENED SHOPAHOLIC

35 "The United States consumes more energy, water, paper . . ." Cited by the nonprofit Center for a New American Dream and its president, Betsy Taylor, who says that America's growing obsession with acquisition is taking a heavy toll on the environment.

49 "John W. Stamm, dean of the School of Dentistry . . ." Cited by UPI, "NewsTrack: Science," May 2002.

54 "In a recent report . . ." World Wildlife Foundation, *Cork Screwed: Environmental and Economic Impacts of the Cork Stoppers Market,* 2006. The report is online at www.wwf.org.uk/filelibrary/pdf/corkscrewed.pdf.

CHAPTER 3: SAY "AAAAH!"

65 "When New York health researchers noticed . . ." The study is the Long Island Breast Cancer Study Project. It's online at http://epi.grants.cancer.gov/LIBCSP/; Breast Cancer Fund.

76 "Recent studies add to the worries . . ." National Cancer Institute, 1994; cited by Environmental Working Group at www.ewg.org/node/26022.

76 "Another study . . ." *International Journal of Cancer*, February, 2001.

77 ". . . a recent study found DBP and other toxic phthalates . . ." U.S. Centers for

Disease Control and Prevention, *Third National Report on Human Exposure to Environmental Chemicals,* 2005.

81 "The Institute for Agriculture and Trade Policy (IATP) found . . ." The study was jointly conducted by the IATP and the Sierra Club and announced in 2002. The press release is online at http://lists.iatp.org/listarchive/archive.cfm?id=62741.

82 "PAHs form when fat from meat drips onto the charcoal." Cited by the American Cancer Society, http://tinyurl.com/2gaof4.

82 "Researchers have identified seventeen different HCAs . . ." National Cancer Institute, cited at www.cancer.gov/cancertopics/factsheet/Risk/heterocyclic-amines.

90 "The Environmental Working Group concluded . . ." Cited at http://environment.about.com/od/earthtalkcolumns/a/chlorine.htm. EWG also said that some eleven hundred other smaller water systems across the country also tested positive for high levels of contaminants.

CHAPTER 4: LIVING (AND WORKING) SPACES

102 "In laboratory studies, some prenatal exposure to PBDEs . . ." Washington State Department of Ecology, Brominated Flame Retardants in Consumer Products: Environmental and Public Health Concerns, www.ecy.wa.gov/programs/swfa/mrw/pdf/Presentations/AnnBlakeBrominatedFlameRetardants.pdf

105 "Fortunately, the Michigan-based Ecology Center . . ." *Toxic at Any Speed: Chemicals in Cars and the Need for Safe Alternatives,* 2005. It's online at www.ecocenter.org/dust/ToxicAtAnySpeed.pdf.

109 "The federally funded National Institute of Environmental Health Sciences (NIEHS) concluded . . ." *EMF: Electric and Magnetic Fields Associated with the Use of Electric Power,* 2002. It's online at www.niehs.nih.gov/health/scied/documents/emf2002.pdf.

120 "NASA researcher Bill Wolverton first reported in 1984 . . ." *Plants Clean Air and Water for Indoor Environments,* 2007. It's online at http://ntrs.nasa.gov/archive/nasa/casi.ntrs.nasa.gov/20080003913_2008001482.pdf.

120 "And a 1994 German study reported . . ." "Detoxification of Formaldehyde by the Spider Plant and by Soybean Cell Suspension Cultures," in *Plant Physiology*, 1994.

121 "Scientists who studied the issue found that the dishwasher uses only half the energy . . ." University of Bonn, *A European Comparison of Cleaning Dishes by Hand.* It's online at www.landtechnik.uni-bonn.de/ifl_research/ht_1/EEDAL_03_ManualDishwashing.pdf

CHAPTER 5: PHANTOM LOADS AND ENERGY SUCKERS

142 "According to the *American Journal of Public Health*, every 10 percent increase . . ." Cited in Jeanne S. Ringel and William N. Evans, "Cigarette Taxes and Smoking During Pregnancy," *American Journal of Public Health*, November 2001. It's online at www.ajph.org/cgi/content/full/91/11/1851.

144 "A study predicts sixty-five billion dollars in U.S. wind investment . . ." Cited by Emerging Energy Research, Cambridge, Massachusetts, www.emerging-energy.com.

144 "According to researcher Wallace Erickson . . ." Cited in Wallace P. Erickson et al., "Collision Mortality of Local and Migrant Birds at a Large-Scale Wind-Power Development on Buffalo Ridge, Minnesota," *Wildlife Society Bulletin*, Autumn 2002.

150 "Cornell University researcher David Pimentel . . ." Cited at www.news.cornell.edu/stories/July05/ethanol.toocostly.ssl.html.

151 "Addison Bain, a National Aeronautics and Space Administration (NASA) researcher, investigated the *Hindenburg* crash . . ." Cited by *ESD Journal*, www.esdjournal.com/articles/hindenbrg/hindburg.htm. Also, "Hydrogen Exonerated in Hindenburg Disaster," *National Hydrogen Association News*, www.hydrogen-association.org/newsletter/ad22zepp.htm. A refutation of Addison Bain's theories is at http://spot.colorado.edu/~dziadeck/zf/LZ129fire.pdf.

CHAPTER 6: GREEN THREADS

179 "Singer Bono, along with his wife, Ali Hewson, . . ." According to www.edunonline.com, "EDUN, in conjunction with Verité, is in the process of developing and implementing a multi-phase Corporate Social Responsibility strategy that will include monitoring of all EDUN suppliers, Performance Improvement Planning, capability training at the corporate and factory levels, and stakeholder engagement." The clothes are currently made in Peru, Tunisia, Kenya, India, Mauritius, and Madagascar.

CHAPTER 7: THE WHOLE KID AND CABOODLE

190 "The issue received considerable attention following an article . . ." Robert F. Kennedy Jr., "Deadly Immunity: Robert F. Kennedy Jr. Investigates the Government Cover-up of a Mercury/Autism Scandal," *Rolling Stone*, June 20, 2005. It's posted online at www.rollingstone.com/politics/story/7395411/deadly_immunity.

194 "Each day, more than a million children ages five and under take in unsafe levels of pesticides from food consumed at home." Environmental Working Group, *How 'Bout Them Apples?*, 1999. It's online at www.ewg.org/reports/apples.

194 "In 2004, 28 percent of all beverages consumed in the United States were carbonated soft drinks . . ." Cited by the American Beverage Association.

197 "The packaged meals derive two-thirds of their calories from fat and sugar." The Center for Science in the Public Interest, "Back to School: Lunchbox Makeovers–Ten Tips for Packing a Healthy Lunch for Kids." It's online at www.cspinet.org/new/school_lunch.html. The story concludes, "Making your own healthy alternative is as easy as packing low-fat crackers, low-fat lunch meat, a

piece of fruit and a box of 100% juice in your child's lunch box (at the very least, use the lower-fat Lunchables)."

201 "Some researchers believe, however, that the threat of nut allergies . . ." Meredith Broussard, "Everyone's Gone Nuts,'" *Harper's*, January 2008. Posted online at www.harpers.org/media/slideshow/annot/2008-01/index.html.

203 "Rock star Bono and others recently tried to call attention . . ." The finger-snapping ads date back to 2004, and were developed in Britain by Abbott Mead Vickers ad agency as part of the "Make Poverty History" campaign. In addition to Bono, celebrities included model Kate Moss and (in France) singer Johnny Hallyday and actress Juliette Binoche.

205 "Yes, the toxic legacy can start at birth." Environmental Working Group, *Body Burden: The Pollution in Newborns*, July 2005. Online at http://archive.ewg.org/reports/bodyburden2/execsumm.php. The findings are based on tests of samples of umbilical-cord blood taken by the American Red Cross from ten babies born in 2004 in different part of the United States. The most prevalent chemicals found in the newborns were mercury, fire retardants, pesticides, and the Teflon chemical PFOA.

206 "According to a federal report . . ." Centers for Disease Control and Prevention, *Blood and Hair Mercury Levels in Young Children and Women of Childbearing Age—United States 1999*. Cited in *Morbidity and Mortality Weekly Report*, March 2, 2001. The report is online at www.cdc.gov/mmwR/preview/mmwrhtml/mm5008a2.htm.

CHAPTER 8: FEELING THE HEAT

217 "The Pew Center on Global Climate Change suggests . . ." *Observed Impacts of Global Climate Change in the U.S.*, 2004. Online at www.pewclimate.org/docUploads/final_ObsImpact.pdf.

227 "The U.S. Forest Service has studied the use of trees for carbon sequestration . . ." David J. Nowak, et al., "Oxygen Production by Urban Trees in the United States," *Arboriculture and Urban Forestry* (International Society of Arboriculture), 2007. The report is posted online at www.nrs.fs.fed.us/pubs/jrnl/2007/nrs_2007_nowak_001.pdf.

232 "In 2006, the UN released a report entitled *Livestock's Long Shadow* . . . The agency says the 18 percent of all greenhouse gas emissions from livestock is more than from transportation. United Nations report, *Livestock's Long Shadow*, 2006. Online at www.virtualcentre.org/en/library/key_pub/longshad/A0701E00.htm.

CHAPTER 9: OPEN ROAD

244 ". . . no less than a dozen car companies now offer partial zero emission vehicles (PZEV) . . ." A complete list of PZEV cars available in the United States (not necessarily in all states) is posted by the state of California at www.driveclean.ca.gov/en/gv/vsearch/cleansearch_result.asp.

255 "Major airports rank among the top ten industrial air polluters . . ." You can purchase clean energy offsets to make up for the environmental impact of your air travel. The two-page booklet *Flying Green: How to Protect the Environment and Travel Responsibly* is downloadable from the Tufts Climate Initiative at www.tufts.edu/tie/tci/carbonoffsets/TCI-offset-handout.htm.

258 "As *E* describes it . . ." Cited in Jim Motavalli, "Taking the Natural Path," *E – The Environmental Magazine*, July/August 2002. Posted online at www.emagazine.com/view/?351&src=.

CHAPTER 10: TECHNICALLY SPEAKING

267 ". . . Apple has come under fire from the Silicon Valley Toxics Coalition . . ." The group says that more than 80 percent of U.S. e-waste is exported to impoverished countries. Its website at www.etoxics.org/site/PageNavigator/StoryofStuff offers excerpts from the very entertaining video about planned obsolescence, *The Story of Stuff.* Your perspective on consumer goods will never be the same again!

271 "Nikki and David Goldbeck's book *Choose to Reuse* . . ." The Goldbecks maintain a website at www.healthyhighways.com. Their company Ceres Press offers many of their books on environmental topics and vegetarian cooking.

274 "Recent research shows that many common electronic devices qualify as hazardous materials . . ." A PowerPoint presentation based on this University of Florida study, *Leaching of Hazardous Chemicals from Discarded Electronics,* is at http://swix.ws/cd/epa_dvr-atl/content/pdf_atl/vann.pdf.

278 "A study of 420,000 cell phone users in Denmark . . ." The study was reported in the *Journal of the National Cancer Institute* in 2006 and is summarized at www.tinyurl.com/2ftzmh.

278 "However, Swedish scientist Lennart Hardell has made a link between brain cancer and older analog cell phones . . ." *European Journal of Cancer Prevention*, June 2002. Published online at http://tinyurl.com/2dpqqb.

278 "The results of a 2003 study do not bode well for habitual chatterers . . ." The study, "Rat Brain Damage from Mobile Phone Use," was published in *Environmental Health Perspectives*, June 2003. It's online at www.ehponline.org/docs/2003/111-7/ss.html.

CONTRIBUTORS

Principal authors:

DOUG MOSS is the founder (1990), publisher, and executive editor of *E – The Environmental Magazine* and the creator and editor of "EarthTalk." Doug is involved in *E*'s day-to-day editorial planning, fundraising, circulation, and marketing efforts and in special projects such as this and *E*'s previous book, *Green Living: The* E Magazine *Handbook for Living Lightly on the Earth* (Plume, 2005). He also edits and distributes each week's "EarthTalk" column to its sixteen hundred media partners and coordinates the promotional efforts to place the column in its growing list of newspapers, magazines, and websites. Prior to founding *E*, he cofounded the *Animals' Agenda* (1979), a bimonthly animal protection magazine, serving as an editor and its first publisher until 1988. He also founded (in 1979) and owns Douglas Forms, a supplier of printing to magazines, nonprofit organizations, and other businesses. He lives in Westport, Connecticut, with his wife, Deborah Kamlani, an *E* cofounder, and their two sons.

RODDY SCHEER, *E*'s contributing editor and formerly the webmaster for emagazine.com, contributes articles to *E Magazine* as well as to *E*'s weekly e-newsletter, *Our Planet*. He is also the primary researcher and writer of the internationally syndicated "EarthTalk" column on which this book is based. Beyond his duties at *E*, Roddy is a regular contributor at *Seattle* magazine and has written for a wide range of other regional and national publications. He is also a working professional photographer, with his stock collection of over ten thousand nature, wildlife, outdoor, and travel images represented by the Danita Delimont agency. Samples of his writing and photography work are available on his website, roddyscheer.com. He lives with his wife and two children in Seattle.

Editors:

JIM MOTAVALLI is the former editor of *E* and author or editor of five books: *Forward Drive: The Race to Build "Clean Cars" for the Future* (Sierra Club Books, 2000), *Breaking Gridlock: In Search of Transportation That Works* (Sierra Club Books, 2001), *Feeling the Heat: Reports from the Frontlines of Climate Change* (Routledge, 2004); *Green Living: The* E Magazine *Handbook for Living Lightly on the Earth* (Plume); and *Naked in the Woods: Joseph Knowles and the Legacy of Frontier Fakery* (Da Capo, 2008). He regularly appears on radio and television to discuss *E* articles and also hosts his own radio show on WPKN-FM in Connecticut. He is a regular featured columnist in the Environmental Defense Fund's newsletter, contributes to thedailygreen.com blog and writes regularly on automotive and environmental subjects for the *New York Times*. He lives in Fairfield, Connecticut, with his wife and two daughters.

BRITA BELLI is the editor of *E*. In addition to editing and writing, she appears on *E*'s behalf on numerous radio and television programs coast to coast, including the Comcast Network, the Regional News Network, and Connecticut's News Channel 8. She maintains the blog www.playitgreen.com on environmental intiatives in sports. Prior to joining *E*, Brita was the arts and entertainment editor of Connecticut's *Fairfield County Weekly* newspaper, where she won numerous awards for her writing from the Association of Alternative Newsweeklies, New England Press Association, and Connecticut Society of Professional Journalists. Her stories have been featured in the books *Notes from the Underground: The Most Outrageous Stories from the Alternative Press* and *Best AltWeekly Writing and Design 2006*. She lives in Fairfield, Connecticut, with her husband and young daughter.

Illustrator:

CHRIS MURPHY has been illustrating for more than fifteen years for such clients as the *New York Times*, *Boston Globe*, *Forbes,* and, of course, *E*, as well as advertising agencies, greeting card companies, book publishers, and more. You can visit his website at www.cmurph.com.

The editors wish to gratefully acknowledge the editorial assistance of two *E Magazine* interns, Kelly N. Hughes of Quinnipiac University and Samantha Grasso of the University of Connecticut.

ABOUT *E*

E – The Environmental Magazine debuted in 1990 while the world was celebrating the twentieth anniversary of Earth Day, yet reeling from a series of environmental shocks, including the *Exxon Valdez* oil spill, "greenhouse summers," fires in Yellowstone Park, and medical waste washing up on eastern shores. In the time since, *E* has established itself as the leading independent environmental journal.

Edited for the general reader but also in sufficient depth to appeal to the dedicated activist, *E* is a clearinghouse of information, news, commentary, and resources on environmental issues. *E* was founded and is published by Connecticut residents Doug Moss and Deborah Kamlani. *E* is a project of the nonprofit Earth Action Network, which also owns and manages the environmental website emagazine.com, where an extensive archive of *E* stories is maintained. *E* also produces the weekly *Our Planet* newsletter (which appears at www.emagazine.com and is e-mailed to thirty-eight thousand subscribers) and, of course, the syndicated question-and-answer column "EarthTalk," which appears weekly in hundreds of newspapers, magazines, and websites throughout the United States, Canada, and around the world.

E's two previous books are *Feeling the Heat: Dispatches from the Frontlines of Climate Change* (Routledge, 2004) and *Green Living: The* E *Magazine Handbook for Living Lightly on the Earth* (Plume, 2005).

E covers everything environmental—from recycling to rain forests and from the global village to our own backyards—and reports on all the key and emerging issues, providing substantial contact information so readers can investigate topics further or plug into hands-on efforts.

E also follows the activities and campaigns of a broad spectrum of environmental organizations and provides information on a range of lifestyle topics—food, health, travel, house and home, personal finance, consumer product trends—as they relate to environmental quality.

E has drawn considerable recognition since its launch, garnering a dozen awards and citations for its style and content. *E* won the Independent Press Award for Best New Magazine upon its founding in 1990 and in 2003 received three Independent Press Award nominations and won in the category of Best Science/Environment Coverage. In 1999 and again in 2007, the Population Institute awarded *E* a Global Media Award for Excellence in Population Reporting.

Many *E* articles can be found in newspapers and other magazines, primarily by arrangement with featurewell.com, and are posted on many environmental websites. Reprints of *E* articles are used widely by environmental organizations in waging their public and media education campaigns.

Subscriptions are $29.95 per year (six bimonthly issues) and can be ordered by writing: *E Magazine*, P.O. Box 2047, Marion, OH 43306 U.S.A.; by calling: (800) 967-6572; or by visiting *E* online at www.emagazine.com.

INDEX